MW00682713

Better to Travel Hopefully, Than to Arrive

The Memoirs of Robert George Collett

Robert George Collett

**Grosvenor House
Publishing Limited**

This book is published by
Grosvenor House Publishing Ltd
Link House
140 The Broadway, Tolworth, Surrey, KT6 7HT.
www.grosvenorhousepublishing.co.uk

A CIP record for this book
is available from the British Library

ISBN 978-1-83975-547-7

This is a record of the days of my life from 1934 to 2020 from Baildon, (Wallington, Whitgift), Trojan's, REME, CAAE, De Havilland's, AGI, Sandal Precision, Vauxhall Motors, Hunting Engineering and the RAE. Plus, two marriages and a series of holidays from the middle of the 20th century, through the start of the 21st in Great Britain, Ireland and Northern Europe.

'It is Better to Travel Hopefully, Than to Arrive'.

A quote by Robert Louis Stevenson

'This is what I have been doing all my life'.

My epitaph

And this is what Fred and myself have also been doing for the past 45 years.

'It is one thing to have an idea, it is another to have the technical and executive ability to give it flesh; it is still another to have the tenacity of purpose to drive it through to success, unshaken in confidence in the face of discouraging opposition.'

Roxbee Cox, Lord Kings Norton, 1947

'Publish or Perish'.

Roger Moss, 2008

Dedication

To my ever-loving, tolerant parents, family and friends who have helped and supported me through all of my life so far.

Stoney Stanton, Leicestershire
May 2020

Introduction

My initial reason for writing these memoirs is that, just after the first Foresters Drive 'Mob' reunion on Sunday 7 August 2011, Miles, myself and one or two others were talking about genealogy and the past etc, when I made that statement; *'well, I don't know anything really about my parents past as we never talked much about it, I think they met at a Tennis Club near my mother's home in Wrexham.'*

Where upon Miles said, *'We don't know anything about your past either,'* so that was it, down went the gauntlet!

I started writing in pencil, easy to rub out mistakes etc and I felt that it was more personal being handwritten, I still think so! It was initially, only for the family and close friends and also, only the first 45 years or so, as the children would by then, be of an age to comprehend. However, it spiralled out of control into a book for all those interested, and a further 40 years have been added to bring it virtually to date.

Up until about 2015, it was written in the garden room more or less on my own. Then came a lay off for some years until 2018, when I transferred to the lounge where it was warmer, and Jill could take part especially helping to sort out the holiday photographs and correcting grammatical and spelling errors. This has helped a lot since Jill's death, the long, wet winter of 2019–20 and the arrival of COVID-19. I've also been helped by keeping in touch over the years with old friends from most of the firms I worked for, school, social and even as far back as Baildon i.e. 80 years! Plus, a good memory and a comprehensive set of records and photographs etc.

All of it in time, more or less as it happened. There are many things that even my closest friends do not know about and it has helped me to understand myself a little better by writing it down, especially when it has led on to other thoughts where I have been able to make useful, I hope, comments and observations. There are still gaps that will probably never be filled, like where are the missing persons of the 'Mob' etc and what became of them? I'd love to know! I never realised what hard work it is. The writing is the easy bit. It's all the research, checking dates, sifting through photographs etc, that takes most of the time: never mind it's good exercise for the brain.

Will the next generation read it? Will they find anything of interest, in a world of 'virtual reality?' Who knows? Well, I hope I haven't left anyone out, if you are not mentioned in the text you will probably be on one of the photographs! And to coin a corny phrase, I hope this gives you as much pleasure reading it as I've had in writing it. And now I must get back to work on the E11.

R. Collett
May 2020

Contents

The MOB 1950–55

Boys

*	Robert Collett	
	Richard Symes	
	Martin Fenny	
	Peter Gill	
*	Chris Hayward	Local
	Burt Burgess	Wallington
	Willie Wilson	
*	Colin Shields	
*	Bobby Hamilton	
	Colin Wright	
	David Oatley	
	David Grey	
*	Barry Roach	
	Bob Cherry	Whitgift
	Ernie Evans	
*	Geoff Gunter	
*	Mike Lane	
*	John Lane	
*	Mick Davis	
*	Clive Heasman	
	Clive Tomsett	
	Gordon Cornell	
*	Bob Durrant	Beddington
	Geoff Brown	
	Bob Skelton	
	Jim Dale	

Girls

	Maureen Rendall	
*	Carol Roach	
	Pat Curry	Local
	Janet Chafee	
*	Anita Chafee	
	Cynthia Harrison	
	Stella Hanson	
	Iris Johnson	Other
	Louise Burken	
	Maureen Hayes	

* Those who attended the first Foresters drive 'SEAT' reunion at John and Lyn Lane's house, Swingate Farm on Sunday 7 August 2011.

Chapter One

My Earliest Days in Baildon and The War From 1934 to 1945

30 Ferncliffe Drive, Baildon, Shipley, Yorks.

Born in Shipley Yorkshire, named Robert George, I lived at the above address; with Father – Robert Cecil, Mother – Violet Authry, and sister Phyllis Ann Collett. My earliest memories were of the milkman arriving on his horse and cart most days, whereupon my mother would go out to the road with an assortment of jugs etc, we never saw milk in bottles or heard of pasteurisation until we arrived after the war in Wallington, Surrey.

Visits frequently to Leeds to visit three aged aunts of my mother's resulted in exciting rides on trams that ran between Bradford and Leeds in those days. One day, I was startled to see a Zeppelin pass nearly overhead. It turned out later to be the Graf Zeppelin, on a flight in May 1936. Some say it was also on a spying mission of Britain's northern industrial towns as a prelude to the Second World War!

About this time, I was sent at the age of five, to a private Kindergarten school called 'Miss Kays', where I learned, by rote, my twelve times tables, the alphabet – A to Z and various other life's requirements; like good manners etc.

One incident pre-war that stands out was a trip in my father's Morris Ten car to Bolton Abbey. At least that is where we were supposed to go, however we had only got as far as Ilkley Moor when the engine boiled over on one of the steep hill climbs and steam filled the car, so we all bailed out and sat by the roadside wondering what to do next. Fortunately, a car drew up and it turned out to be Mr Bottomley, our neighbour who lived opposite. He took mother and us children home whilst my father sorted out the car! He wasn't technical in anyway, so I supposed he had it removed to some garage for repair. Later, when the war started and petrol rationing etc, he sold it, or at least it disappeared! It would be 20 years before he had another car, around 1960. We lived on an unmade road which terminated at a vertical cliff drop of about 100 feet on Baildon Bank which had in years past, been quarried out.

These quarries ran from our road, south towards Shipley Glen and Saltaire for about a mile. At one point some quarter of a mile from us they had turned 90 degrees and excavated a huge square hole in the ground, considerably deeper than the cliff quarries. It must have been about 150 feet deep. So much so, that at the bottom, the sun only reached in Midsummer. The whole ecosystem existed with plants, trees and even a small stream that emerged and disappeared into what seemed solid rock. I mentioned all this because in our childhood minds, we had a desire to explore. One of our main objectives was to get down into this pit! Needless to say, we never did! They had built a large wall on the cliffside, to presumably keep people out, but half of it had fallen down in our time.

It was generally considered a dare to climb this wall and of course I had to have a go. I always think of this as my first brush with death, I chose a bad part of the wall. It was okay initially, but as I got higher it curved outwards into an overhang and for a while I could not go forwards or backwards and I froze for some minutes, hanging on by my fingertips and feet. Eventually, after a supreme effort, I managed to continue and reached the top, a much chastened and wiser boy! I never climbed that wall again, even the easier end that everybody climbed.

Baildon was a superb area for children growing up in the days before television, computer games, etc. When children had to entertain themselves, using initiative, ability and imagination etc.

In addition to the quarries and cliffs, I have mentioned, there was the bank, 45-degree rock strewn and running the opposite way i.e. north from Ferncliffe Drive for three-quarters of a mile to Baildon itself. Then there was Hope Hill to the west of us, it was 920 feet high, almost a mountain and it formed part of Baildon Moor, which included Ilkley Moor, part of the all-inclusive, Rombolds Moor. Going beyond the aforementioned cliff was Shipley Glen, a rugged moorland with a steep drop down to the River Aire and Saltaire. There is a funicular cable operated railway which is over 100 years old and runs from the Baildon Top down to Saltaire.

Beyond Baildon, on the north side was the moor which stretched to Ilkley and beyond with some incredibly old miners' cottages, called 'Moorside', which were occupied during my time but were all knocked down in the 1960s, as was the centre of Baildon, Croydon, Bath and many other places. I call this 'the decade of vandalism', when many lovely buildings etc were pulled down in the name of progress! High rise flats predominated and look what happened to them. Our immediate area was dominated by the cliffs and bank at the end of our road and the adjacent quarry.

When I left the kindergarten school, I went to Sandals Road School at the bottom of the bank with most of the other children in the area. The journey was alright in the summer but in winter it was another thing altogether! It had a single zig zag path starting at the top which stopped halfway down. This half had proper steps and a handrail but after this it was just a slope with no handrail. In ice and snow, it was lethal, difficult enough on this upper half, almost impossible on the lower. Many were the times; I've had to go down on my backside trying not to lose control and slither off the path onto the snow packed grass alongside! It would not be allowed now – health and safety, etc.

I enjoyed my time at this school during the war years, gas masks, air raid drills, bomb shelters etc, and one or two incidents spring to mind. One, I remember when I was climbing up a guttering drainpipe onto the school roof to retrieve a ball that had lodged up there. I took particular trouble to get it back as it belonged to a girl I fancied! It was my first emotional encounter with the opposite sex.

Unfortunately, climbing on the school roof was regarded as a punishable offence, but the authorities didn't know who it was. The headmaster read it out at morning prayers, demanding the culprit be revealed, threatening the whole school with detention but nobody spoke up and the whole affair evaporated! I think the reason for this mass silence was my popularity at this time, as I had freed almost all of the school children during a game of 'Chain He'! Now, 'Chain He' is not to be confused with 'Off Ground He', it is a game of catch, where one person is the catcher and when he catches any of the others they have to link hands and form a chain until one of the free persons can touch the chain to free them all again! Well, on this occasion, I think everybody involved, which was most of the school, was in the enormous chain and I was the only person who could free them! Which I managed to do, sacrificing my own freedom in the process! I was always a good runner!

Of course, during most of this time the war was on, my father was in the Home Guard, not too old to fight in this one, but he had served in the Navy in the First World War in 1918, on minesweepers in the English Channel. He also was in a reserved occupation – Income Tax Inspector! Say no more! I always viewed his bayonet with enthusiasm, it was the old-fashioned bladed type, just like a sword, but it was kept out of sight most of the time, sensibly.

About this time, I was introduced to Meccano, a wonderful device, not a toy, which I took to like a duck to water, giving me, I think, an engineering insight that was to last the rest of my life, the desire to make things. If only I had known that the Scott Motorcycle manufacturers were based in Saltaire, Shipley, just a few miles away!

Also, at this time, we made several visits to Wallington, Surrey and I remember the steam train journeys from Bradford station to

London via Chesterfield where we used to look out for the leaning church tower, a well-known landmark. We went down to see my grandparents on my father's side, who lived at 12 Ingleby Way where we were eventually to end up after the war. Several times we heard the sirens go and had to get out of bed and crouch under the stairs or go to the Anderson shelter in the garden if the raid was severe. whilst here, we felt first-hand, the effects of the war; bomb noise, blackouts etc, whereas in Yorkshire, on the edge of the moors, we saw and heard very little, although both Bradford and Leeds were bombed.

At the start of the flying bombs V1 and V2, my dad's sister, Aunt Dorothy and two younger children; Christine and Martin were evacuated and came to stay with us. Christine, going to my school temporarily, whereas Martin was too young. He was a bloody nuisance and would spend his time tormenting my sister, Phyllis until one day she snapped and hit him over the head with her doll, which was of the type that had a soft body and a hard head! Oh well! It laid him out cold and of course panic ensued. My mother and Aunt Dorothy rushing him upstairs to bed, phoning the doctor etc. Pandemonium rained for a while. The final result was that he recovered, fortunately without serious injury. Aunt Dorothy insisted on punishment for Phyllis but my mother and the rest of us figured he had got his just desserts! So, Phil got off with a light smacking and he never teased her again!

One night, we had an earth tremor and I remember waking up to hear the crockery and loose objects rattling about. I initially thought it was the bombs on Bradford or somewhere. Two incidents rendered me extremely unpopular with our next-door neighbour; the first when Barry Wingfield received a brand-new football (you must remember at this time during the war such things were unobtainable) I don't know how he got it but there it was and he naturally treasured it to the envy of the other children! On this occasion, he was playing with it in his driveway when it came over the low fence into ours. I saw what happened and taking a swipe at it, attempting to return it to him. Unfortunately, my kick was misdirected, and it smashed through their kitchen window and horror of horrors, his mother and grandmother were having their midday meal and all the glass debris had fallen onto the table. Needless to say, I got whacked for it, lost my pocket money to pay for it and the only amusing bit about it, if it can be called amusing, was seeing Granny's face peering out through the hole! The second incident with this wretched football was a few weeks later when foolishly, Barry loaned it to me for a kick around and I decided to go down the road and walk along the top of the cliff at Baildon Bank, kicking the ball in front of me. Well, you can imagine what happened, it was of course inevitable, down it went all the way to the bottom, almost outside the school.

Well, I started off down, when to my horror I saw a van pull up and a group of men got out, picked up the ball and disappeared in the direction of Shipley. For this, I got a sound thrashing from my father. He used the cane he kept at my mother's insistence, But I am not sure how the loss of the ball was resolved, I assumed my father had to pay for it!

Just to compound the issue, around this time, a group of lads and myself managed to set the field at the back of our garden alight. The situation was retrieved by Barry's uncle Doug, my father and several other adults, after much beating and water was used. My popularity was at an all-time low, especially as it was found that Uncle Doug's army trousers were burnt! He was home on leave!

I'm afraid, I also set the hedge at the back of Hitchiner's house, backing onto the quarry alight, this was accidental as I dropped the match during an illegal smoking session, however Mr Hitchiner was very reasonable, taking the view that, 'boys will be boys.'

Now, this aforementioned quarry backed right up to the houses opposite us on the other side of the road and anybody climbing over their back fence was literally within four or five feet of a huge shear drop to the bottom. Sure enough, one day we came very close to disaster!

Alan Whittaker lived almost opposite us and his house backed onto the quarry, we were all playing out there when he somehow slipped but he fell onto a sloping area about six to eight feet long, but as he tried to scramble up the loose scree started to slide, taking him with it. We couldn't reach him and every time he tried to move, he slid even closer to the edge. Somebody fortunately had gone for help and his father arrived and told him not to move and he laid down full length gripping the upper edge of the slope, his feet now being within reach of Alan's hands.

'Now,' he said, 'grab my feet and climb up me which he did.'

It was a close call, we had all been repeatedly warned by our parents but after this, all of us were much more careful.

David Berry, my oldest friend reminded me of this incident circa 2003, when Jill and I visited him and his wife Pauline in Otley where he now lives. On another idiotic occasion, David and I rolled a boulder down Baildon Bank and it hit the wall of a cottage at the bottom. Immediately, a woman appeared and started up the cliff after us, needless to say it was a futile attempt, but it brought home to us the danger of the situation, we were learning slowly!

The winters during these years were very severe and usually lasted from about October to May. Snow from November to April. It was made worse as we lived on the edge of the moors at an altitude of about 600 feet. On several occasions, I have seen my father open the front door to face a complete blockage of drifted snow overnight; we would have to dig our way out! Also, being on an unmade road, there were plenty of puddles in the gutters etc and when these froze, plenty of skating took place. Also, there was a large lake, now drained, which was alongside Baildon Green, which froze every winter and large numbers of the local inhabitants would enjoy themselves in winter sports, it was surreal, all lit up, with hot dogs etc 'on tap'!

I had a push bike at this time but most of the kids seemed to venture around on foot. David Berry and I spent many evenings walking with his father over the moors; on to Hope Hill and Tong Park, covering considerable distances especially in summer with the long evenings and wonderful views. We were also in the Cubs – Junior Scouts – and on several occasions, we went on chases over the moors. On one, David and I went first leaving signs on whatever, rocks, walls, etc with chalk and the rest of the pack followed trying to catch us before we got home. We went such a long way, became exhausted eventually and decided to hide in a haystack but we were discovered and so they won!

My mother's brother, Uncle John often came to stay. He was what would now be described as having 'learning difficulties' but he was a good sport and used to come with us to the Glen Royal cinema in Shipley on Saturdays for the children's films – mainly westerns and comedy. We also used to go to the cinema in Northgate, Baildon in the evenings to the adult cinema but the only film I can remember seeing was *Hold That Ghost* featuring Abbott and Costello, who were an American comedy duo on the lines of Laurel and Hardy.

My father, working of course for the civil service, was suddenly posted to Croydon in 1944 at the same time as the 'Doodlebugs' i.e. flying bombs, previously mentioned, started to rain down on the South of England! I remembered my mother naturally being very upset and of course she had the additional burden of running the house etc and looking after his kids.

There was a lot to do in the days before fridges, washing machines, dishwashers, toasters, liquidisers and central heating, etc. My mother cooked on a range that was coke fired and coke was rationed like everything else. With no central heating, at night you rushed up to bed and jumped in curling up into a ball to keep warm.

She did have a 'Charlady' that came in, I think, two mornings a week but one of them, Julia was caught stealing jewellery and dismissed pronto! It was a long way to the shops in Baildon, about a mile each way, no buses or taxis etc, no petrol! My mother walked it, pulling a trolley behind her with the shopping in it. Finally, the war in Europe ended, VE Day and a large bonfire was lit on the moor with street parties and celebrations taking place everywhere. Father returned and arrangements were made for us to go and live in Wallington, Surrey at my grandfather's house; we left in the late summer of 1945 before VJ day, as I remember.

1. I returned to Baildon several times thereafter, first in 1956 with Richard Symes.
2. With my father when we stayed in a B&B in Ilkley in 1988.
3. And again when we established contact with David Barry and Pearl in circa 1990.
4. 1993 with Jill, David and I had our walk around all the old haunts!
5. Also, with Jill and Phyllis circa 2006/07 when we again had a B&B in Ilkley.
6. Circa 2008 when returning from holiday in Scotland with the Lanes to visit D. Berry in Otley.
7. And in 2019, after Jill's death with Miles, Sarah, Peter, Paul and Jenny to scatter Phyllis's ashes in the garden of 30 Ferncliffe Drive, over Baildon bank at the end of the road and Shipley Glen where our B&B was. We also visited the D. Berry and Pearl in Otley.

First years of my life - 1934 to 1945.

Chapter Two

My Time at Whitgift – 1946 to 1950

When we first arrived in Wallington, I think I must have missed the start of the new autumn term as I seemed to have some weeks in which to amuse myself. The Wallington area, at this time, was almost as interesting to a small boy as was Baildon; the points of interest were:

Croydon airport, just at the end of our road which, at that time, had hangers that were full of the remnants of war, like aircraft that had never been flown, live ammunition and no end of other supplies!

Roundshaw Park and the adjacent allotments also at the end of Ingleby Way.

Lordsbury Field, which lay between Hamilton Way and Stratton Avenue, bordering on to Buckingham Way; good for aeromodelling activities.

Woodcote House and extensive grounds which also included Fenny's Hollow and the surrounding woodland area right through to Foresters Drive and the seat. Fenny's Hollow was also used for aeromodelling, mainly control line, as it was not as big as Lordsbury Field and was surrounded by high trees. The house had been requisitioned during the war for the development work on the De Havilland Mosquito, so I understand.

Pampisford Road, this was our cycle route to school each day and had many old, exceptionally large houses that had also been requisitioned and were now empty!

Beddington Park, the attraction here was the lake on which we used to sail our model boats, main problem was the leeches,

11

that would attach themselves to your legs if you weren't careful!

So, I cycled around exploring the new environment. One day when returning from a visit to my Aunt Louie who lived in Coulsdon, about five miles away, I was stopped by a man, who after asking me various questions, turned out to be a school inspector! Next thing I know is that I am attending the local, what is now comprehensive school – Bandon Hill, similar to Sandals Road in Baildon.

I didn't stay long, my father moved me to Collingwood Preparatory School from where I sat the entrance examination for Whitgift. I remember the first interview with the headmaster, a Mr Ingham. He found difficulty in understanding what I was saying due to my very broad Yorkshire accent, my father acted as interpreter! By some strange chance I passed, what would now be called the '11 plus', I suppose.

At this time, I formed my first southern friendship with Richard Symes by throwing an almond nut at him, while seated in the nut tree which was in our next doors garden. We hit it off straight away as he was also from the North of England, i.e. Lancashire – Formby to be exact. We were friends for life until his untimely death in 1989, but more of that later.

He had also taken the Whitgift entrance exam and passed so we started the autumn term of 1946 together. The starting day was also a parent's day and sports day combined. I was informed that I was in 'Junior house A' and would be expected to run in the handicap race.

Well, I was handicapped at the front because:

  a) I was a new boy with no previous record and,
  b) I was small for my age!

But I fooled them! What they didn't know was my running ability, learned on the Yorkshire Moors! I went off like a rocket and kept it up for the entire race, although I could hear pounding feet coming up behind me fast and just as I crossed the line a huge boy overtook me, but it was too late!

Mr Symes was so pleased he gave me half a crown 2/6 – 12 and a half pence in modern money. The trouble with this was that they now expected me to run and take part in all sports the house was involved in; my view was that having won one, I had demonstrated my ability and that was that! This attitude set me on a collision course with the housemaster and other keen sporty types, so much so that Herbert Gurr, myself and other anti-sport types used to crouch down at the back of the room hoping not to be seen at house meetings when they were looking for volunteers.

Also, at this time, events were taking place at Croydon Airport that were to have a profound effect on our lives. There were all kinds of people including us lads, who were breaking into the hangers on Croydon Airport and thieving all kinds of stuff, such as military clothes, food packages, tools etc. In fact, we pinched some rope and a block and tackle assembly and rigged that up in Roundshore Park from a high tree over what appeared to be a bomb crater and had a great time whizzing down the rope, at high speed, to a crash landing on the other side! Also, in addition, climbing onto the roof of the hangars and then jumping off and sliding down the canvas awnings until driven off by the arrival of the police!

On another lunatic expedition, we acquired some live tracer bullet ammunition, all this was just lying around just after the war! Amazing. We wondered what to do with it and some bright spark, seeing a bonfire on the adjacent allotments – threw it on and all hell broke loose, as shrapnel started flying everywhere, endangering everybody, including some pedestrians walking along the Foresters Drive pavement! We made ourselves scarce very quickly. We used to squeeze through the railings and sail our model boats on the fire hydrant pond until chased off, usually by two Laurel and Hardy type policemen, they were too fat to get through the gap amid much laughter!

But all this hilarity was about to end in disaster when an older youth, not of our gang, fired a stolen very pistol off on Lordsbury Field one foggy day when an aircraft was attempting to land on Croydon airport. It had a narrow escape, nearly hitting the roofs of the surrounding houses. Well, the resulting police investigation gathered up enough people for both the juvenile and senior courts to be involved. I don't think any of our lot were prosecuted, we were all at most in our early teens, but we all got a good telling off, also from our parents.

However back at Whitgift, to get me out of sporting events every Wednesday and Saturday afternoons, my mother was co-opted to write sick notes to get me off sometimes; then two things happened:

1. 'Claude', this was the name of our housemaster; I was now in Smiths House; suddenly said one day. "there are certain boys in this house who are not pulling their weight and as a punishment, they will not be allowed to participate in any games in future", what a laugh! Herbert and I could not believe what we were hearing; played right into our hands.
2. Masters Robinson –woodwork and potter – art must have convinced the headmaster – Marler, that there were certain boys who were not cut out for sport and in consequence a hobby section was arranged under their supervision for Wednesday afternoon.

The aeromodelling group comprised Richard Symes, Ernie Evans, Bob Cherry, Chris Olsen, Stuart Uwins, Ebbut, several others and me; we flew our planes on big side which was a south eastern slope featuring two or three rugby pitches.

We were already making and flying model aeroplanes on Lordsbury Field and Fenny's Hollow, as mentioned before, when

Mr Fenney, with two of his work colleagues decided to join us on a regular weekly basis. They were experts in scale model aircraft, both control line and free flight, plus early radio control. Their models were beautifully made and they had made a ducted fan version of a doodlebug, German VI – brilliant!

There were two incidents of note on Lordsbury Field whilst flying our planes, one involved Stuart Unwin's model Horsa glider, which he was launching using a bungee system of catapults, it came out of the sky from an enormous loop and hit Richard on the back of the neck and he went down poleaxed; he was standing right next to me. We all thought the worst but after a moment or two he recovered but was considerably shaken, a bit higher and he would probably not have recovered – dangerous! Similarly, one of the lads' free flight models got away and crashed into the upper bedroom window of a house in Hamilton Way, near where Maureen Rendall used to live. Whereupon an irate mother appeared saying that the broken glass from the smashed window had landed on her baby's cot! She took a bit of pacifying.

I did get caught for a Rugby match once but proved to be so useless by avoiding the ball and kicking it away whenever possible. Similarly, in the cricket matches when asked to field in the slips, I turned away when the batsman struck the ball, having seen another boy taking a ball in the face with painful results! One redeeming feature was bowling; I could throw a reasonably straight ball and being left-handed most batsman found this disconcerting, nevertheless not my scene.

Symes and I, and sometimes Barry Roach, would cycle to school every day via the longer southern route around Croydon Aerodrome, making up and telling each other adventure stories to pass the time away, also sometimes exploring some of the old deserted requisitioned houses that existed at the time just after the war. There was a huge place called Kendra Hall in Pampisford Road that had been empty. We broke into it several times finding lots of interesting things to do.

On one memorable occasion, Symes and I were exploring around the outside of the building in the kitchen garden, when, on looking around a wall, I saw two policemen obviously looking for us! As they had not seen us, we decided to do a runner and, on the word, 'go', we set off at high speed across the back garden, towards the rear wall, that backed onto the reservoir. They shouted and set off after us, we had just enough lead to be able to scramble over the high brick wall and fall over the other side almost on top of a courting couple lying on the grass! They were not best pleased of course, especially when the police arrived seconds later, but we didn't stop to argue, ran like hell, leapt onto our bicycles and away!

On another, rather more serious occasion, accompanied by another schoolmate, H.R. Roberts, whose idea it was, we broke into a large house in South Croydon, late one winter evening. We did all the usual things, exploring, where we found a billiard room with table etc, and proceeded to have a fencing match with the cues! Whilst we were in one of the upper storey bedrooms, we were accosted by two shifty looking men, they asked us what we were doing here and several of the questions regarding our school. They said they knew the Headmaster and would not report us if we left and said nothing!

Later in the week, H.R. Roberts, turned up with the Croydon Advertiser in which there was an article about two escaped convicts in the area! Wow, what a narrow escape, we said nothing to anybody not even our parents. It didn't seem to curb our enthusiasm as we continued to explore empty places in Downside Road, Purley, where racing driver John Surtees was to live eventually and the Oaks, a massive house in the Oaks Park near Carshalton, subsequently pulled down in the vandalism of the 1960s. Here we discovered the bell room in the servant's quarters and had a grand old time, ringing all until the inevitable law arrived.

This cycling to school resulted in much punishment for me in the form of canings by the Prefect's for riding in the school grounds, which was forbidden. My problem was being late in the mornings for main assembly at 9:00am, I would have to scoot and was inevitably caught by a Prefect Patrol. This meant that one went before the Prefect's Court on Friday lunchtime and was sentenced to 'six of the best', the maximum allowed for regular offenders. You were offered an alternative detention and lines, but it was best to get it over and done with quickly!

Another sport to be dreaded was swimming, this I particularly disliked, and cannot swim even today; remember in those days the baths weren't heated and although I had almost learnt to swim at Sandown in the Isle of Wight in the hot summer of 1947, I particularly hated getting water in my eyes, ears and mouth, particularly chlorinated water! Awful.

Now, Whitgift did not have its own swimming pool then, fortunately, but occasionally we were dragged by some keen Master to the Croydon Baths. I used to hide in the cubicles as much and for as long as possible this was even worse than rugby and cricket!

One of the few sports I like, if you can call it a sport was 'British Bulldog'. This, I think is a 'peculiarity' particular to British Public Schools, the idea is to split the class into two halves, line them up facing one another, whereupon they rush at each other and fight. The idea is to get through the opposing line and after a given time of fighting has lapsed, the winners are the side with the greatest number of breakthroughs. Marvellous! I know this sounds a bit odd on my part, but I did like the Gym, swinging on the ropes, climbing on the bars, vaulting over the vaulting horse, etc. I was quite fit and energetic in those days, but I didn't want to do it out of school hours, I wanted to make and fly model aeroplanes, which was my main hobby at that time; before motorcycles took over. Some of our most enjoyable moments were watching the aircraft come and go from Croydon Aerodrome as it was then called. Reference here to the series of books called *Croydon Airport*, a series published in four volumes by the Sutton library, of which I have copies!

We would do this whilst sitting in the grass, smoking our homemade pipes, made out of bored out cotton reels with a hollow bamboo stalk pushed in the side and full up with dried out Raspberry leaves; we tried out several things but decided Raspberry leaves tasted the best! On one occasion, one of the lads had an earwig crawl out of his pipe, so he just pushed it back in again and forever afterwards it was claimed we were smoking crushed earwigs!

We experienced two crashes during my time here, the first was when a Bristol Freighter was landing, having just gone over our heads; when it touched down, the starboard undercarriage leg just folded up and the aircraft slid sideways and skidded along the grass. There were no hard runways of any length at Croydon. The second crash was much more serious, it was on taking off, an Aerovan high wing, twin engines, monoplane, which was flying away from us fortunately towards the Purley way. It seemed to struggle to get off and suddenly turned over to port and buried itself into the ground on open fields near Pampisford Road. The pilot and co-pilot were killed but I am not sure about the passengers as they were a load of pigeon's apparently!

Following our enthusiasm for aeroplanes, helped by living just alongside the Croydon airport, Richard and I joined the Combined Cadet Force (CCF) – air cadet section under the command of Mr Robinson, the previously mentioned woodwork master. This involved a parade after school, I think for an hour every Tuesday evening; there were visits to Kenley Aerodrome on Wednesday afternoons and also at weekends occasionally.

On one such visit to Kenley, Richard and I were taken on a flight to the Isle of Wight in an Avro Anson, it was my first time in an aeroplane and we had to manually wind the undercarriage up and down after take-off and before landing; shows how primitive it all was in those days; this, the pilot had to do on his own normally!

We also had a week's camp, once a year and I remember us going to Cranwell in Lincolnshire, home of the RAF, where we had a great time firing .5" calibre Browning machine guns and flying in North American Harvard and Percival Prentice single engine, low wing aeroplanes, which were the RAF training aircraft at that time, we were allowed to take the controls for a short time and do gentle turns!

Meanwhile, back at home we were up to no good, as usual, making explosives. Somebody had discovered that a mixture of magnesium powder, obtainable from photographic shops for flash photography; sulphur and potassium permanganate, obtainable from the chemist for medicinal purposes was, once lit, highly explosive! So, the homemade, handheld cannon arrived, made of a length of steel tube, flat at one end with small hole for a length of 'Jetex' fuse, and strapped onto a suitably carved wooden handle. Which when ignited usually with a fag, was quite capable of blowing out streetlamps, putting holes in people's fences, killing anybody or any other lunatic ideas. We would probably have been arrested for terrorism today!

We started off firing ball bearings and/or cycle Cotter pins etc, but soon progressed to casting our own lead bullets, usually in Kevin O'Callaghan's garden shed. We drilled holes in two house bricks with a filling hole, clamped them together and poured the molten lead in, simple. Unfortunately, we did suffer from lead poisoning, due to inadequate ventilation and very unpleasant and dangerous it is too, lead is cumulative, say no more!

Our second weapon was the homemade hand grenade, this was a mixture of sulphur and potassium permanganate, poured in between a nut and two large bolts which, when thrown and collided with a hard surface, like the tiles on the roof of Woodcote House, became an impact explosive, stripping the bolt threads, hurling the three pieces and a load of tiles in all directions. It was bloody dangerous, and on one occasion could have had very serious consequences. This took place almost outside Chris Hayward's house in Stratton Avenue, when a large group of us had gathered with our push bikes presumably to await Chris's arrival; when someone, I know not who, threw the grenade which landed just alongside Colin Wright! However just at that moment he had slid forward off the saddle and onto the crossbar, whereupon a piece of shrapnel went through the saddle removing the seat, the springs and everything else! I can see it now, I was right alongside of him and we were both very shaken, we never did find out who threw it.

Our third explosive device was an Acetaline bomb which comprised of a Tate and Lyle syrup tin, containing a medicine bottle with Acetaline and water plus a Jetex fuse, all packed with earth, which was dropped into a previously dug hole, usually under a tree. Wait a few seconds, light the fuse, and run! Willie Wilson was our specialist, he could blow a small tree or large bush right out of the ground, roots and all! Really spectacular, as it was not just the explosion but the ball of fire that accompanied it!

Another, once a year event, was Founders Day when the whole school marched in columns right through Croydon to the parish church at Reeves Corner, meeting up with the Whitgift Middle School of which Bob Durrant was a member. Every section of the school was represented, starting with the Drum and Fife Band, Navy, Army, Air Force Cadets, Scouts, Masters and finally the rest of the boys in descending order, the first forms coming up at the rear. It was a morning do and we all got the rest of the day off. Now this reminds me of a visit to the school by Eric Barker, a popular radio comedian of the day, who later turned up in Ealing comedies and *Carry On* films and was an 'OW' – Old Whitgiftian. He gave us a talk, then to everybody's surprise, even the Headmaster, I think, he asked if we could all have the afternoon off! He got a big cheer for this and an embarrassed Headmaster!

The History master, a Mr Edge, stands out as a character; he had served in the war with the 'Chindits', General Wingate's forces in Burma fighting the Japanese and it was not reckoned to be a nice business! Naturally as you would expect he was a disciplinarian, but fair; on one occasion after Symes had been particularly difficult, he said, "I am going to give you regular beatings for a week," so each time we appeared and before Richard could do anything he was caned! Mind you, he only kept it up for about three days, Symes got the message.

Edge was a forward-thinking man and used to encourage us boys to speak in public. He would set aside half of the history lesson sometimes and select a boy to stand up in front of the class and talk about any subject he thought fit. On my turn, I gave a lecture on astronomy which I was and still am very keen on; so much so, that he asked me to do it on several occasions. Richard lectured on archaeology; he was very keen on visiting old castles.

Symes got whacked for rapping on his desk when one of the master's made the remark, "it's not often that fortune knocks on your door". Ebbut for running his model aeroplane engine under his desk during a rowdy English lesson with Mr Kennedy who struggled to keep order; and various other boys for shorting the lights by putting silver paper between the contacts.

Around this time somebody had the idea of upstaging our efforts with a super cannon and firing it at Woodcote House from across Fenny's Hollow, a distance about half a mile! So, we all put our pocket money together to buy the necessary Magnesium mixture and a large, approximately two-and-a-half-inch diameter by four feet long steel tube was obtained and suitably modified. Of course, a special bullet had to be cast, and this was done between two bricks in O'Callaghan's shed. The whole contraption was embedded in earth on the side of the hollow that faced Woodcote House and observers were posted around the house to record a hit! Previously some argument ensued as to who was best qualified to determine the elevation and azimuth, and David Grey had won! A large crowd of supporters turned up, nearly all of the 'Mob' and others. A huge explosion with the sound reverberating across Wallington, plenty of smoke and nothing else, all the observers recorded 'no hit'. A disaster for whom most of the blame fell on Grey. However, a week or two later an article appeared in the

'Wallington and Carshalton Times' newspaper from a man who lived in the Drive complaining of a hole in his roof and a lead ball in his attic and further accusing the Croydon airport authorities of not insisting that pilots of low flying aircraft on the approach to landing should wind in their trailing Aerials earlier!

Well, it all made sense, we immediately went round to have a look trying not to be too conspicuous and ascertained that his home was in direct line with Woodcote House and the cannon, so the shot was a success of sorts, it obviously had gone right over Woodcote House, it must have gone almost a mile! Mr Fenny knew about it and told us all to shut up.

Next up, but not necessarily in this order, as my memory is some 70 years on was Richard Symes rocket! This was entirely his own idea so the rest of us were just innocent bystanders! We knew he was working on something secret and eventually he invited us to the great event on Lordsbury Field, one winter evening. He had filled a steel tube with what he called a slow burning explosive; upended it and placed it into another close-fitting steel tube about two inches in diameter, as its launcher. By the time everything was ready, and it was firmly affixed in the ground, vertically – note that it was dark also! There was some apprehension, but Symes ignored this and lit the fuse. I must say it worked remarkably well and with a trail of sparks it disappeared into the gloom! Then we all froze, including myself, if only he had inclined it slightly, it landed with a great thud in the grass after what seemed like an eternity, which of course it could have well been for someone! Finally, to wrap up the explosive phase of our careers we used to strap homemade bangers onto arrows and shoot them from homemade bows from our back garden onto the roofs of the houses in the Newlands, a road that backed onto us in which Burgess and Snowball lived, until the neighbours complained.

At Whitgift, one of the more serious episodes occurred during a geography lesson with Parsons. Now, Parsons was not a man to trifle with, he had only to look at the class for silence to be complete. On this occasion, an argument between Richard and a boy called Robinson had started, I can't remember what it was about and it doesn't matter anyway, but all of a sudden it flared up into violence and Robinson who was sitting in the desk in front of Richard suddenly turned round and with sheath knife stabbed two or three holes in Richards Atlas, narrowly missing his hands! Parsons immediately took them both to see the headmaster, which if I remember correctly, resulted in a severe lecture to them both and a caning for Robinson. It just shows how easy accidents can happen, Robinson wasn't a bad chap really, but short tempered perhaps. Anyway, they made it up, but Robinson later left us, I had the impression that his parents could not manage the fees, which were £25 per term at that time. I remember this figure as my mother had to sell some of her shares to pay for it and used to complain when my school reports came in at the end of term with remarks like 'could do better', 'does not try', etc but to her and my father, I owe an unpayable debt for their unstinting support, which I feel was not repaid until I obtained a First-Class Diploma in Aeronautical Engineering from the College of Aeronautical and Automobile Engineering, Chelsea in 1958. And on the strength of this, went to work for the De Havilland aircraft firm as a development engineer, but more of this later.

There was a chap at school at this time called Spiv Ward, he was older than us, a bit of a braggard, and one day we got to arguing about the relative strength of our gangs, the net result was a fight to take place in our woods next to Woodcote House, which we declared to be our territory!

Well, on the day selected, he turned up with his gang at the seat of Foresters Drive and Richard and I were the only two of our lot to appear! There must have been about 12 of them, all armed to the teeth with sticks, catapults, bow and arrows and in some cases, including Spiv himself, real swords! They said they would give us 10 minutes start, so we took off fast. We just headed off home it was the only thing to do. The following days examination of our Woodland camp etc showed utter devastation they had destroyed everything. Needless to say, Richard and I upbraided the rest of our gang – 'where were you?'

On another occasion, we did get a full turn out and the battle for Woodcote house took place. When the Beddington Mob, led by Rex Duke and David Hancock, a cousin of Iris's, although I didn't know it at the time, appeared one day in force and started to attack us with air pistols, homemade wooden swords and catapults etc. we had a fine old scrap, blokes rolling about on the ground wrestling and punching each other. Geoff Gunter had his own battle with another thug, firing air pistols at one another at point blank range, each yelping with pain as they took a hit! Richard and I were up on the top floor of the house hurling down whatever we could lay our hands on, when a large stone from a catapult shattered the window and frame bringing down some of the brickwork also on us. The battle raged on and off for most of the afternoon with groups of people from one side or the other chasing or being chased all over the grounds and adjacent woods. The final result was inconclusive, as it finally petered out with the arrival of teatime and a general lack of interest and recriminations of 'your swords bigger than mine'. On another occasion, we found a load of gas masks and spent some considerable time sitting in a closed shed with a smoking bonfire; playing cards whilst we tested them out! I told you we were crazy; we could hardly see the cards!

I pinched a large flat door from somewhere in the kitchen area and took it home, set it up on trestles in Dad's garage, where we all used it to play Ping Pong; Table Tennis to you! Whilst all this was going on, we went in for a bout of tunnel exploration which started, I think, with the discovery of a tunnel by the River in Carshalton Park near the War Memorial Hospital I was later to attend! Following this, there was an extensive set of tunnel systems under the Coulsdon Downs, that had been used by the army ordinance for storage during the war; accessible once we had broken in. The tunnel system we had the most fun with was right alongside the Plough pub, at the bottom of Queenswood Avenue where the Lane's used to live. There was talk of it linking up with the Priory in Beddington Park. It had been bricked up for years due to the danger of collapse, as it ran under Sandhills. It took us some time to chisel our way in as any noise was liable to promote interest from the nearby pubs and shops. Eventually, after several days of after school working, we made a hole big enough to squeeze through but found out that we had gone through the roof therefore requiring a drop of about six feet or so! Once in, you could stand up and exploration showed that it divided into two tunnels, one larger than the other but both blocked eventually by an extensive roof collapse that even we could not penetrate!

Schoolfriend, Ballard's father, had a yard and Wharf in South Croydon and we made an oil drum boat with scaffold planks and sailed it until one day it sank, fortunately without incident. I remember one afternoon towards the end of my school days while sitting in class, Geography I think, when I began to lose vision with bright zig zag flashes occurring, I was naturally frightened and went to lie down in the sick bay. It's all passed off in about half an hour, apart from a dull headache which disappeared after a night's sleep. I later discovered that I had had a migraine from which I have suffered on and off all my life; fortunately, mine are not serious, they can be quite debilitating, I believe. I have no real idea what causes them, I can have two or three in quick succession all within a week and then not have another for two years or more. They can be triggered by stress, alcohol, sickness, Illness, colds, etc, but I've never been able to confirm anything definite; apparently, they are quite common with many sufferers.

So, I came to the end of my school days taking O levels in Maths, Physics, Chemistry, Biology, Metalwork, English and Geography. I wished I had paid more attention now but 'wiser after the event'! On to Trojan Ltd!

12, Ingleby Way, Wallington, Surrey - 1946 to 1966.

Chapter Three

Apprentice at Trojan Limited – 1950 to 1953

Trojan Limited was a very interesting company situated at the end of Trojan Way just off the Purley Way, Whaddon. It was founded by Leslie Hounsfield around the early 20th century and produced automobiles featuring a novel two stroke engine, he designed himself. It was a square four-cylinder, two split singles side by side with their crankshafts set at 180 degrees. This engine went on until the 1930s, when it was redesigned by J.B. Perret into the same split singles augmented with a pair of charging cylinders at 90 degrees. Enough said!

When I joined the firm in the autumn of 1950, nearly 17 years of age they were producing light, one-ton vans, with the Perret engine; mini motors which fitted above the rear wheel of an ordinary push bike and drove on the rear tyre by means of a roller! They also manufactured hydraulic hand pumps and a helical oil grooving machine, plus various sub-contracts.

The vans were sold to Wall's ice cream, Brooke Bond Tea, Harrods of London and anybody else who could be persuaded to buy!

The mini motors came in two sizes 50cc and 75cc single cylinder two strokes, they were the brainchild of an eccentric Italian who used to visit the factory on occasions. I say eccentric, although I never spoke to him, he was far above a mere apprentice! He just seemed odd, waving his arms about, talking nineteen to the dozen, a big hairy face, full of whiskers. He was, however, clever because we were the first of the cycle motors as they are known, and they sold in large numbers. Some said they saved the firm from bankruptcy as van sales were poor. Later on, when the idea caught on, various other makes appeared like Power Pack, Cyclemaster, BSA winged wheel, Scott Cy-auto and one or two others.

I started in the general stores at 30/- i.e. £1.50 in modern money, per 48-hour week, 8:00am to 5:30pm and then moved to the tool stores; This was the general rule of things, the idea being to give the newcomer a general overview of all the components items etc that go into a production factory manufacturing their own designs. From here we went to the simplest of machines, the drills and learned how to sharpen them properly etc, and then onto the general machine shop i.e. Lathes, Capstans, Mills, Grinders, Autos, etc. It was on the grinding section, operating a surface grinder that I made the big mistake of not switching on the magnetic table before starting the cut, the result was a noise like a machine gun and all of the components left the table at high speed when the grinding wheel hit them. Luckily, I had put the guard up so no one got hurt! I was not popular as can be imagined.

Whilst we were attending to the practical side of things during the day, we attended night school at the Croydon Polytechnic, three nights a week from 7:00 'til 9:00pm to study for an Ordinary National Certificate (ONC) in mechanical engineering for three years, then a further two years study for Higher National Certificate (HNC), it was a long day!

Later on, of course, they went on to a 'one day release' with possibly only one night, but not in my time.

We studied: – Engineering Drawing – Projection, etc Maths, including Calculus for HNC, Applied Mechanics, Engineering Science and Machine Shop Practice.

It was an extremely comprehensive course and with the hands-on practical apprenticeship experience was capable of turning out competent engineers and designers. In other words, a person who could imagine an idea, set it down on paper, details and all; make it, assemble it and further develop it into working entity! I feel that the University route that followed these wonderful apprenticeships with all of the teachings I have described, fell short in the over-dependence on academia and examination results to the almost exclusion of the practical. After all, an engineer or artisan, is a man who makes things, but this seems to have been forgotten!

It seems to me that a large number of today's 2012 graduates have not had their imagination stimulated and have little idea how to make things: the tool capability and usage and particularly the material section requirements and their treatments. Also, a lot of the 'fun' has been taken out of it because, in my day for example, if you were say designing a spring, you sat down with pen, paper, log tables etc, reference books and slide rule and worked it out! Then the slide rule and log tables etc were replaced by first mechanical, then electronic calculators and eventually the pen, paper and reference books were no longer needed as the computer arrived! Now all you need to do is type in the sizes, forces, loads etc – input data and press the key and out comes your spring to do the job!

It obviously saves the employer time, it does in seconds what previously took hours, but I feel that it has taken a lot of the interest and glamour out of engineering and is probably the cause of the decline in popularity, such that with the poor pay, the brighter types head for the city or other white-collar jobs. Youngsters today don't seem to want to get their hands dirty either.

However, to get back to my apprenticeship days, as I previously said, Trojans mini motor drove by a roller onto the back wheel of a push bike. This, as you can imagine, was okay in the dry weather but slipped like hell in the wet causing also much tyre wear. Richard, quick off the mark as usual, was first to do something about it. He took the 50cc two stroke engine and bolted it under the frame in a horizontal altitude similar to the Italian Motor GUZZI and the all-chain transmission drove through a rear wheel hub assembly comprising large and small sprockets on this countershaft. There was no clutch, one paddled the machine forward using the decompressor then whisked it onto throttle; the same handlebar lever did both functions. I followed his idea almost immediately and put the engine on the front down tube resembling a Sloper Panther motorcycle and had the chain drive down to the normal pedal shaft. Thus, using the already made bearing countershaft arrangement; the rest of the operations were the same as Richard's machine. We used these to go to work for some considerable time until we got our first true motorcycles. Mine was Corgi, which was used by parachutists in the war, they folded up so that they could be fitted into a tube and dropped separately by parachute. Eventually, I sold it and bought a 250cc New Imperial, which was coil ignition, dependent on a good battery. I tried to sort out the wiring but with my limited knowledge at the time, it beat me, so I swapped it for a 1928 James 197cc with a Villiers engine, flywheel magneto, no reliance on batteries for ignition! This eventually was replaced by the only new motorcycle I have ever had, a 98cc James Comet, the money lent by my father.

One memorable ride I had on this, was a trip up the A12 to Felixstowe, a distance of approximately 120 miles to visit and stay with my parents who were holidaying there. It was a long and dispiriting ride at a mere 45mph maximum most of the time, it was only 120 miles and seemed to take all day! This was the last time I went on holiday with my folks; the army intervened and then after that I went with various motor cycling mates. Meanwhile, the Mob originally started in Carlton Avenue, around Richard Symes house, number 31, where he spent most evenings after work, messing about with motorcycles, accompanied by lots of helpers or hinderers, as you like it!

He took over all the garage and most of the garden with his rubbish until his long-suffering mother, brother John and also long-suffering neighbours, who had noisy motorcycles roaring past their houses every night; called it a day!

I don't know who suggested the seat on Foresters Drive but it was an ideal place for large gatherings with the woods behind and Croydon Airport in front and no houses for at least a quarter of a mile each way; extending from Roundshaw Park to Colin Wright's bungalow in Plough Lane, a distance of at least one mile. It was an ideal place to meet, chat and ride our bikes as there was virtually no television, computer games or mobile phones etc. and the expresso coffee shop had not arrived!

"It was the place to be," as Maureen Rendell said!

There was a good selection of British motorcycles in order of superiority, Mick Davis's Triumph Speed Twin; Clive Heaseman's BSA Gold Star, Symes's Triumph T100, Colin Shields DOT and later on my Arial 600cc square four and several others from the Beddington Mob, that includes Bob Durant's Royal Enfield Bullet, Mike Lane's Velo and John Lanes 250 BSA. Our star performer to impress the girls was Colin Shields (sadly no longer with us, as I write this in 2020) who could stand on his saddle and ride down the road with no hands on the bars! Just like the Royal signals display team! Nobody else achieved this. He was also able, on his Trials BSA Bantam, to climb lamp posts; to explain, he would stand on the footrests ride up to a lamp post or telegraph pole, yank the front wheel in the air and place it on the post, keeping the bike in this position by slipping the clutch; or back wheel if on the grass, for as long as was necessary! Crazy but clever!

However, most of the serious speed work was done on the Purley way, the 'no speed limit bypass' between Purley and Whaddon, where you could do over 100 mph in reasonable safety! Best I ever achieved was a 95-mph speedo reading on my Ariel square four but Richard, I think got over 100 mph on his Tiger 100, all this with no crash hats or goggles! Stupid.

Naturally, some romantic attachments were made during this time i.e. Rendel and Burgess, Pat Curry and Peter Gill, Heywood and Janet Chapman, Symes and Shields vied for Janet Chaffee's hand,

I was never quite sure who won, if either! And Anita and I, she was my first girlfriend. I've lost track of quite a few – refer to frontispiece – but one or two ended up abroad i.e. Maureen Randell in America, although still in touch; Janet Chaffee in Australia, still in touch; Stella Hanson, Australia, racing greyhounds, so I understand, and Jim Dale, Australia, now departed! It all disintegrated eventually as people went into national service, got married, lost interest or got on with their lives; after running for about five years 1950 to 1955.

Getting back to Trojans, several things occur to me, one was the incredible noise of the place, Health and Safety would deem it injurious to health. The reason for this was twofold; the main cause was the line shafting that drove virtually every machine. Only the gear cutters, if I remember correctly, had their own independent driving motors. Secondly, the noise of the automatic production lathes autos; the machines themselves were quiet, but the long metal bars rattling around in their support tubes some 10 or 12 feet in length made a lot of noise. Another racket, although short lived, was the giant press that stamped out the van roofs, it stood as high as a house and when it came down the whole place shook! On several occasions, we had power cuts when everything would stop and then the maintenance crew would rush about to get our own diesel-powered generators, of which there were two, operating.

Many of our mad schemes, like adding milk crate sidecars to our mini motor push bikes, or one to one geared trials push bike sprocket for peddling through bracken and blackberry bushes, were made during this time as 'homers', which was a term given to home jobs, not firms work!

There was, once a month, an apprentice meeting with the management, usually chaired by the Managing Director, a Mr Williams, with the Chief Engineer, Mr Perrett, Production Manager a Mr Tovell and several other senior figures. It was very interesting, and it gave us a chance to ask them questions and air any grievances we might have. They took place in the canteen, which was alongside the Trojan museum, which housed some remarkably interesting cars, vans etc, that the firm produced in its early days. I often wish that I had had the two-stroke knowledge I possess today, because I would have dearly loved to questioned Mr Perrett, particularly about his design problems, decision's etc that resulted in the van two-stroke engine etc, but alas it was not to be.

At this time, David Grey decided he wished to own a motorcycle and for some odd reason decided to ask me to accompany him, I can't imagine why, as I knew little or nothing about real motorcycles i.e. 125cc or above. So we set off, down to Bill Lavender's shop at Wallington Green, where David became interested in a 250cc BSA but could not make up his mind so Bill and I went off up the road to his other shop to look at a very interesting NUT. 750cc [Newcastle Upon Tyne] Next thing we knew was a white-faced Grey at the door, saying the bike was on fire! Old Bill moved like lightning with us following managing with the help of a fire extinguisher to beat out the flames. It transpired that in trying to start it, it had backfired through the carburettor, setting the float chamber alight and very nearly the fuel tank! Grey bought the bike.

On one dark and foggy evening, when some of the Mob were gathered together in Stratton Avenue, I was 'whining' air pistol pellets off the road, they made a sort of singing sound as they disappeared into the gloom. After several shots, one of them suddenly stopped short with a thud! And next second a cop appeared, so I took off down the road with him in hot pursuit amid shouts of laughter from my fellow hooligans. I sped down the pavement for about 100 yards with him gaining on me, when I came alongside the Fenny's Garden fence, which having considerable impetus, I took in one almighty bound, landing on the lawn and flowerbeds. Then I double back up the garden, came out towards the house and hid under the bench in the garden shed.

Meanwhile the cop was opening the garden gates and looking in the garden generally, all the noise had aroused Mr Fenny who came out of the back door for a discussion with the policeman and net result being that he said he would go for a torch! As soon as he opened the door, he saw me crouching under the bench, realised something was up, collected the torch and went out to divert the cop's attention, whilst I made my escape through the side gate! Another narrow escape – some you win some you lose – thanks Mr Fenny.

Two amusing incidents spring to mind, regarding the mini motor; the apprentices, as can be imagined, were a lively bunch of lads, full of fun and always ready for a joke! Well, they/we got it one day. One of the foremen was considered to be a 'nasty piece of work' to coin a phrase and he had upset some of the apprentices, so they extended the high-tension lead, disconnected it from the sparking plug and pushed it up through the holes in the saddle of his bike. Everybody hung about at 'going home time' until he came out, got on his bike, pedalled down the bike sheds, then depressed his clutch bringing the motor down on the rear tyre, as was normal practise to start the engine! He leapt off the bike, amid roars of laughter, as the sparks shot up his arse! Even some of the workers themselves had a job to avoid grinning.

Another wheeze the apprentices got up too was extended running of the mini motor at home, illegally. Let me explain, like the vans, the mini motors had ongoing developments and the apprentices were used as test riders and the standard test route was down the A23 to Brighton and back, distance of 100 miles approximately. Well, those that had done it once or twice got fed up with it and the dodge was to ride it home, set it up there, jack up the rear wheel and leave it running for several hours driving the wheel, so the milometer recorded the necessary 100 miles or thereabouts, and then ride back in the late afternoon having done whatever else, with the engine sufficiently run in and the correct milometer readings!

Coming to the end of my apprenticeship in 1953 reminds me of the Coronation of Queen Elizabeth II, everybody more or less seemed to be glued to the television, people had to share televisions in those days! Richard and I had other ideas. He had a fore and aft 350cc Douglas at this time and we fitted it with one of our milk crate sidecars, which was a tubular frame of the old fashion milk crates with a pair of push bike front forks and a wheel bolted on, with this we proceeded to lap our cross-country circuit in the Woodcote House grounds and Fenny's Hollow; All gone now I'm afraid, it's a housing estate and Fenny's Hollow is called Ambrey Way! Awful. However, to resume, eventually after lots of fun and hairy moments, we returned to my house for running repairs.

Up to now, Richard had been the driver and I the passenger and in spite of some argument, he still insisted on driving, it was his bike after all, so we sallied forth from our drive just in front of a police car, who proceeded to pull us over. Well, they had a field day, no road tax, no insurance, no lights, horn, chain guards, etc, the lot! The whole issue went to court in Sutton, I think and funny as it was, the Justice of the Peace, judging the case was Bob Durrant's father-in-law to be. The milk crate sidecar caused much mirth and Richard got away with several fines and his licence endorsed!

Also, at this time, Colin Wright appeared in court a couple of times for excessive noise. He was picked up whilst riding his push bike with a 10cc Nordic model aeroplane glow plug engine, attached to his handlebars, driving a 12-inch diameter propeller! He claimed that it certainly helped coming up Wallington High Street past all the shops; should have stuck to the back roads!

Bert Burgess and I went into partnership to produce tandems, i.e. two push bikes combined together. One was the normal one behind the other and the other was a side-by-side arrangement with linked steering except that in this case both could steer. All went well for a while until he decided to turn one way and I the other! The resulting pileup caused much amusement. The other version was okay except that the rear person had to do all the hard work, all this eventually was solved by the fitting of a mini motor and this was then used for some time around Wallington by the two of us!

Later on, a milk crate sidecar was fitted so a third person could be carried but a crash at the bottom of Stratton Avenue with Foresters Drive resulted in the whole contraption being wrapped around a tree just outside David Oakley's house.

About this time Richard suddenly decided to sever his apprenticeship and went to do his National Service with the East Surrey Regiment based at Kingston Upon Thames; he was unfortunate not to get in the REME, but they did not have any vacancies at that time.

He also bought a Triumph Tiger 100, 500cc twin cylinder engine with bronze cylinder heads. It was a 100-mph bike and very much upmarket from our previous machines, so on the strength of this I went to sell my puny 98cc James Comet, with the idea of purchasing something similar! After visiting several dealers in, I think, the Brixton area, I was on the way home, somewhat dispirited, when in Streatham I tried once more.

There was nothing suitable within my price range but the showman, a somewhat shifty individual said he had a private Ariel square 4, 1939 vintage of 600cc, which he would be prepared to do a direct swap for. I suppose he thought he could get more for my newish James Comet. Anyway, we retired to his house and the deal was done. When I got it home my father expressed horror at the size of it, saying it was too big for me. He was right of course, as I managed to crash it twice!

The first time was on our holiday to Devon and Cornwall, we got as far as Woodmansterne, where we slid under a car in the wet with a bald front tyre. This was typical of my stupidity or lack of cash, but we started off in the rain, with a heavily loaded bike and Richard on the pillion with all the camping gear, we should have had a sidecar, as in later years – we live and learn, don't we?

As we approached a T-junction, I applied the brakes and the bike laid itself down and the front end went under the wheels of an approaching car. The chap got out and I asked him if he would back off my bike so we could see the damage which was extensive, requiring a new front wheel and forks later; his car wasn't damaged; Richard said it was the most comfortable crash he had ever been in; so it would be for him, he went down the road cushioned by all the camping gear!

Finally, my apprenticeship came to an end, somewhat ignominiously, through a series of unfortunate circumstances! Somebody, I know not who, was caught selling stolen mini motor parts and an investigation was started, various individuals including myself were hauled into the offices and grilled by the management and security. Being young and inexperienced in these matters, I owned up to having some bits and pieces! Which I had been using for experiments with the chain drive bike etc. I was going to convert it to 75cc I remember and had pinched a barrel and head to do this, I had not the faintest idea of selling them for profit but to no avail. Along with several other apprentices, and I believe, some of the staff, we were expelled! Whether criminal proceedings were taken against those individuals who were profiteering or not, I have no idea. It was a harsh lesson, and many felt that it was a little too severe on young apprentices. Unfortunately, Mr Williams, the Managing Director, was on holiday, so the affair was handled by a Mr Monk, who I believe was the chairman. And he was a Methodist! So much for clemency!

Richard avoided it, he had of course gone off into the army, his lucky day. So, with my career in ruins, my parents naturally disappointed, although they said nothing, I departed and immediately applied for National Service, I was lucky to get in to the REME. This in a way was equivalent to two years out in modern terms and gave both me and my long-suffering parents, time to regroup! Without too many awkward questions being asked.

Trojans - 1950 to 1953.

Me with Aerial Square Four - 1953.

Mother on 250cc New Imperial 1934 vintage.

L to R, Barry Roach, Carol Roach, John Symes, Maureen Hayes,
Louise Burken, Peter Gill, Anita Chafee, Janet Chapman and
Chris Hayward. Taken at Shoreham - early 1950's.

Chapter Four

My Days in the Army – September 1953 to September 1955

National Service – REME Army Number: 22919966

I entered the Royal Electrical and Mechanical Engineers (REME) on 17 September 1953 after completing three years of a Mechanical Engineering Apprenticeship at Trojan Ltd, Purley Way, Croydon, Surrey after leaving Whitgift School, Haling Park, South Croydon in 1950.

I was sent to number one training Battalion at Blandford Forum in Dorset for six weeks initial basic training, nothing much of interest happened here except that we were assembled one day and informed that if we did not feel we were suitable for motor vehicle mechanic training, please let them know now as it would save considerable monies not to have to train unsuitable people; no recriminations would be imposed on those who spoke up. From here, I was sent to motor vehicle training camp at Taunton, Somerset, where we spent about four months intensive study both of theory and practice, relating to the overhaul and repair of army motor vehicles. The whole course was extremely good and very interesting to me, especially as we were taught by civilian instructors!

Army discipline was maintained by having to wear uniform, fatigues, fire pickets and guard duties etc, in between times. One of the first things to happen to me was the introduction to the Allen Scythe! We were all assembled on parade when the sergeant, after detailing groups for the cookhouse, officers' mess etc, suddenly asked who could operate an Allen Scythe. Jack Ward, an ex-Trojan apprentice who had joined up the same day as me, pushed me and said, 'you can Bob,'

So, I shouted out, 'I can', even though I had never heard of an Allen Scythe.

Told to stand to one side, I was taken to the machine and expected to start and operate it. Fortunately, whilst I was employing delaying tactics to avoid giving away my ignorance, the sergeant was called away and I eventually fathomed it out!

This proved to be a great 'skive', I was expected to go all around the camp mowing the grass etc and after doing an hour or two of this, I would leave the machine running, out of sight in the woods and skive off to the NAAFI for a game of snooker, cups of tea, etc; it worked because I never got caught! It was during this time that I had my most serious motorcycle accident, which had a somewhat comic sequel!

I went home on a 48-hour pass for the weekend and had arranged to take another chap back with me on the pillion seat of my 1939, 600cc Ariel Square Four. After a day of rollicking around with the lads at home, I picked up this chap from West Croydon Station at around midnight on Sunday. We got as far as Winchester, when descending a downhill section of road, I hit the kerb and crashed headlong into a brick wall! The resulting crash, due to over-tiredness wrote off the bike front end, completely demolishing the front wheel and forks etc, for the second time!! I ended up being taken to hospital for facial stitches, the pillion passenger was intact as I provided a cushion for him and I lost my upper jaw front teeth.

However, upon returning to camp, I visited the army dentist who removed the remaining stumps and prepared to make me a denture plate, which he did with four small teeth in it! I protested that I had only lost two teeth, but he insisted that I have the regulation amount, which is, I believe 16. I explained again that I had only been born with 14; anyway, it is two less than normal! This had been pointed out years ago by our family dentist and it has been recognised by no less an authority than Isaac Asimov, who points out that in 10,000 years' time, most of us will have less teeth, and no appendix! Upon returning home, my mother expressed horror at the four small front teeth and whisked me round to our dentist to be re-fitted with a 'normal' two teeth plate, at her expense.

At camp, life proceeded with lectures on vehicle mechanics for four and a half days a week and military matters for the remainder. The problem was Wednesday and Saturday afternoons which were supposed to be devoted to the GOD sport! For those not interested, it was a chore, but we were lucky to have as it turned out the CO – Commanding Officer who was an aeromodelling enthusiast and so we were allowed to build and fly aircraft during this time. To keep it going, we made sure that at least one or two airplanes were seen in the air during these afternoons and plenty of plans and Balsa wood were strewn around the allotted room.

Eventually, we finished the course and would be sent to permanent postings at home or abroad. Most were being sent to Germany BLOR, Aden and Singapore etc, with a few home postings. It was about this time I began to realise that one could influence one's own affairs to advantage if one did a bit of manipulating and as I had no desire to go abroad, I decided to do something about it.

I discovered that once a fortnight, a dispatch rider came down from the War Office in London with the dispatches, including the postings. I also discovered that no names and numbers were included, these were allocated randomly by a corporal clerk in the camp offices.

He turned out to be an obliging chap and successful negotiations were transacted, i.e. several packets of fags and duly put my name down for Aldershot, which was the nearest to where I lived in Wallington.

It was late afternoon when I arrived in Aldershot (home of the British Army) at the Motor Transport Company, number one training Battalion, Royal Army Service Corps (RASC); and after initial introductions to the other REME lads, we bunked down for the night.

We were a Light Aid Detachment (LAD) group of REME personnel numbering about 20 in all; our CO was a Major (George) a man of about 50 years old. Under him was a Warrant Officer, Sergeant and Lance Corporal, who ran the administration and a further Corporal who was in charge of the electrical section.

The remaining personnel, all Privates (Craftsmen) about six vehicle mechanics, motorcycle mechanic – me, welder, vehicle recovery mechanic, vehicle trim and signwriter, storeman and electrician, tin basher and an armourer. Their job was to repair, adjust and maintain all the vehicles, rolling stock and weapons for the RASC Battalion.

The following morning, I went up before Major 'George' and after explaining who I was etc, he informed me that they did not require a vehicle mechanic but that they had asked the War Office for a welder! Oh calamity! The best laid plans and all that! However, he suddenly added, 'and a motorcycle mechanic'.

Saved! I informed him quickly that I had repaired my own motorcycles and owned three of them. That was it. I was shown the cage in which I was to work on the current army motorcycles, which were the Matchless G3L of 350cc capacity and the most modern looking of the three types due to its teledraulic front forks: the Norton 16H side valve 500cc with girder forks and the BSA M20 side valve 500cc, also with girder forks.

During my time here Graham Valor the new recovery mechanic arrived, he lived in Ipswich and we had several enjoyable rides, he on his 650cc BSA A10 Gold Flash and me on my 600cc Scott Flying Squirrel of 1947 vintage; to visit both his and my parents' homes.

It was on one occasion when returning to camp from such a visit to Wallington with me on his pillion seat that we crashed in Epsom at around midnight. As the bike was damaged, we left it at the Police Station and started walking as the last bus had gone!

We trudged through Leatherhead, Great Bookham to East Clandon along the A426, until we found a comfortable bus shelter, where we dropped off to sleep until the first bus came along. The bike was later recovered by an army truck!

I occasionally used an army bike to go home for the weekend but left it in my parents garage out of sight during the weekend; to use it otherwise, especially in civilian clothes could have been fatal, even going to and fro, riding in proper dispatch riders gear was a risk, being so far off the official test route. Returning to camp in the early hours of Monday morning, in time for the 8am morning parade prior to marching down to the workshops, I would often meet up with a man riding a Norton Dominator, obviously going to work and we would race from Leatherhead to Guildford. He was quicker than me on the straight, but I had him on the bends due to the superior handling of the Scott frame etc. I never ever stopped and spoke to him and I wonder if he thought it unusual, we raced by common consent! and waved goodbye each time. The Scott arrived in early 1954 after I sold the Ariel Square Four to Willie Wilson. My friendship with him suffered a bit after this; mainly I felt due to his mother who thought I had sold him a dud!

The reason for this was that instead of using it as I had, he decided to 'do it up', so he completely dismantled it, the engine anyway, and started to have it re-bored, new pistons, new big ends, main bearings etc, all of which was totally unnecessary really, at vast cost!

She was obviously convinced that all this HAD to be done – so what could I do? Nothing, I tried to explain it, I bought the Scott from Comerford's of Thames Ditton for £70 but it became obvious after a short time that the big ends were knackered, so

I stripped it down using the army facilities and fitted replacement con rods, bushes and rollers etc, purchased from Baxter's of South Norwood where I met John Catchpole, who, at the time, was well known for his competition efforts with the Scott's.

Also, about this time, I came into contact with Eddie Dow, who, with his tuned 500cc Gold Star BSA went on to win the 1955 Clubman's TT (ref photos) in the Isle of Man. He was a Captain in the RASC serving an eight year 'sentence' in the regular army. He was in charge of the dispatch rider section and did 'light' servicing of the bikes, any major overhaul came to me, or if I didn't have the necessary equipment, spares etc. the bikes were sent to the REME works at Borden, Hants. He became a good friend and used to rag me about Two Strokes and the Scott; this was of course, long before MZs started to win world championship races.

Eventually, after suffering this 'abuse' for some time, I challenged him to a race, to show him how quick the Scott was. The net result was that we lined up one evening on the main Barrack Square leading onto the main road with a good crowd of the lads and one or two officers, mates of Eddie's.

The flag dropped, so to speak, and from a standing start, I led him up to about 70–75 miles per hour, whereupon he charged past and roared up the hill toward the Aldershot Hospital. I had previously told him that its strong point was its acceleration, and that top speed was limited to about 85MPH, whereas his machine on TT gearing would probably do about 125MPH.

However, all was reconciled, and he admitted that my performance was a lot better than he had expected. Also, about this time several of the vehicle mechanics were put in for their driving tests, which was also valid for civilian use (I already had the group G motorcycle licence). I was taken out in a three-ton Bedford lorry with crash gears first and second which meant, double de-clutching with a sergeant instructor/examiner. His idea of the test was to drive to a steep hill, stop halfway up, perform a clutch start without rolling backwards, and then, if successful, full speed to Cookham café for tea and cakes, and that was that! Passed! A favourite sport of mine was to perform wheelies across the main workshop floor when empty of vehicles, at that time wheelies were not quite as commonplace as they are now. However, this came to a halt after one attempt went disastrously wrong and I failed to stop... hitting the office door partition, knocking it in and taking the desk, cupboards and Lance corporal who was writing behind it, into a heap at the far end. He was a good lad and agreed not to charge me! Valor got seven days CB (Confined to Barracks) for riding into the cookhouse one lunchtime, egged on by a group of lads who were waiting for their meal; unfortunately, when he arrived at the serving counter, having driven through the mess area, the orderly officer was making an inspection!

The winter on 1954/55 was a cold one and we froze in our stark Barrack room. We had a large stove at the far end of the room which was only of any benefit to those half dozen beds adjacent to it. It was fed with coal from a large bunker also in the room, but the ration was inadequate, so we ended up by burning any wood we could lay our hands on. When we had burnt the fire picket wooden stands, the chairs and anything else we could lay our hands on, we sent out scavenging parties, pinched chairs from the cookhouse and wood from anywhere we could. It all came to a head when Dennis the Painter and some other lads, having had too much to drink in Aldershot one night, saw a 'nice' noticeboard in somebody's garden, whence it disappeared into the fire! Unfortunately, it turned out to belong to a Colonel Llewellyn or somebody important and was reported to the CO. All hell broke loose, but of course, they never found the evidence, but in the course of the investigation the other 'crimes' were discovered. The result was that 'Barrack room damages' was imposed, which resulted, I think, in Half a Crown (2/6d) old money, being docked out of everyone's pay packets for several weeks.

One day, another sergeant arrived from another RASC Battalion riding an immaculate G3L Matchless. It had been assigned to him, he may have been a member of an army demonstration team and he was very proud of it, telling me that he did all of the routine servicing. However, he was obliged by military law to have a regular service by the REME mechanic! This did not please him, as I was his junior in all senses, and he was even more reluctant that I should test ride it. All went well and I completed the service and test rode it, finally parking it on its rear stand outside my cage in the main workshop.

Now Dennis, although unable to ride a motorbike, was always keen to get on one and rev the engine, so, of course, this was a golden opportunity! There he is sitting on his bike, on the stand, revving it up and changing gear so that the back wheel is spinning rapidly, with now a large crowd of onlookers, spurring him on. Well, of course, the inevitable happened, the machine dropped off the stand, shot forwards into a large stanchion that supported the roof, bent the front forks backwards, whereupon everybody, including Dennis disappeared, as if by magic, leaving me to face the outcome.

Fortunately, the Sergeant had gone off to the Sergeants mess or whatever, nobody of any senior rank had witnessed the incident. The forks were solid, what to do? Inspiration arrived; I grabbed some of the guffawing fools and we looped a rope around the stanchion and the forks, and with several hefty pushes backwards, we straightened them by eye until they looked reasonable and did at least move up and down! As the front tyre only had hit the stanchion, there was no visible damage evident.

When he returned, he thanked me curtly and rode off, I breathed a sigh of relief; too soon! He was back in less than five minutes, "what have you done to my bike?"

I expressed surprise, "What are you talking about?"

"It doesn't feel right," and he then proceeded to give it a minute inspection, before eventually riding off finally to roars of laughter.

I didn't see the funny side of it until later. Whilst we are on the subject of Dennis, we had an even more serious incident. He and another chap called Williams, asked me if I would take them out with me on one of my test rides. Like an idiot, I agreed, and we set off, me riding solo on a G3L Matchless and Dennis with Williams on the pillion of an M20 BSA for a 'tour' of the surrounding countryside, intending to end up at Cookham café for tea and cakes. As we were riding along a deserted piece of heathland, Dennis overtook me and the race was on, I caught up and took the lead, probably doing about 60mph and looked over my shoulder to see Williams on the skyline and Dennis and the bike, cartwheeling down the road! Well, the bike was too damaged to continue and worse, Williams required medical treatment, Dennis, needless to say, was unharmed!

We decided to hide Williams and the bike in case the military police or anybody of rank came along, so we found a convenient ditch in the heather, where he crouched down, out of sight. I returned with Dennis on the pillion, dumped him in the paint shop and went off to find the sergeant on duty.

Well, I had to explain it all, leaving out Dennis who had not been missed; however, the sergeant was a decent type and quickly realised that immediate action was required to rescue Williams and the bike, so we returned with the 15cwt truck and bought them back.

Williams was taken to Aldershot General Hospital up the road, I started to repair the bike, which was not too badly damaged, as I remember. When questioned, Williams said they had got into a speed wobble, which is undoubtedly what happened, as it was a straight road but bumpy. After an overnight stay and some patching up, he was discharged and we both went up before 'George'. After a severe telling off, our luck held and we got off Scott-free, I think the fact that Williams had suffered and was due for demob shortly; also, George, who quite frankly, I think did not want the hassle of military proceedings; it could have been a court martial offence on several counts:

1. Damaging army property.
2. Unauthorised use of army property, etc.
3. Not on the official test route.

And, I am sure, the prosecuting council could have added some more!

Whilst we are on the subject of motorcycle escapes, we had another near one, again Dennis and Williams were involved, this time with a sidecar also. We had 'borrowed' Corporal Nobby Clark's Ariel 600cc side valve outfit for a trip to Cookham café one afternoon and were returning at high speed with myself and Dennis on the bike with Williams trapped in the sidecar. The reason I say 'trapped' is that he could not get out, until we got off the bike, as the upper portion hinged towards the bike. Coming over from the brow of a hill with a steep descent on the other side, I was horrified to see a large double decker bus stopped at a bus stop adjacent to the T-junction of the main road at the bottom of the hill. Worse still, when I went for the brakes, the rear locked ok but the front came up to the handlebar without any stopping action at all. So, we went down the hill leaving a black line from the rear tyre with virtually no retarding action from the brakes at all. I could see that the bus driver was about to leave the stop and therefore would have more-or-less completely blocked our path out onto the main road. Standing on the footrests, steering with one hand, I frantically waved him back and he realised that something was wrong, stopped where he was, rested his hands on the steering wheel and watched the proceedings from the safety of his own cab.

Meanwhile, I decided that with the sidecar on the nearside was our only hope was to make a right turn into a side road, the opposite way we desired to go! So, using all of the road we went into a full lock drift, all wheels now sliding sideways, the side of the bus came and went, and we were through with about the thickness of a fag paper between the sidecar wheel and the bus. Upon turning back, I gave the bus driver a sheepish grin, what else! It was entertaining for the passengers! Nobby Clarke was of course, not pleased and I narrowly avoided a charge with some slick talking. The reason for the front brake failure was that Nobby was doing some adjustments/repairs to the brakes and had disconnected the front brake cable; just shows how stupid and naive I was. It seems difficult to understand how we could have got to Cookham café without noticing it!

I got home as much as possible, usually a 48-hour pass at the weekend and also Wednesday afternoon returning early Thursday morning. Talking with some other chaps one day, I discovered that it was possible to continue one's civilian studies if one was close enough to the seat of learning to require only one day out and with of course the permission of the officer in charge i.e. 'George'. So, wheels were put in motion, first a visit to the education officer, he was sympathetic of course and agreed to get in touch with Croydon Polytechnic where I had previously been studying for an (ONC) Ordinary National Certificate in mechanical engineering, during my apprenticeship at Trojan Limited. George was agreeable, so I started a one-day release on a Tuesday every week.

This now meant that I spent very little time at camp, and I would go home on Friday evening and return on Monday morning, leave again Monday evening, return Wednesday morning, go home Wednesday afternoon, half day, usually wasted on sports by those so inclined! return Thursday morning, for the longest spell in camp, two days, one-night Thursday and Fridays. this was a great skive, only one night spent in the barracks, some of the lads were a little jealous I think but nobody said anything.

My mother on the other hand said she thought that when I went in the services, she was going to get rid of me for two years! No such luck I used to get told off, "don't bring your army manners here," she was referring to my 'lying about' on the furniture, armchairs in particular. This was due to the fact that we had no seats etc, everybody used to just lie on their beds with their feet hanging over the end rail. Eventually, this idyllic state of affairs came to an end, George was demobbed; we were all very sorry to see him go, he was a lovely man not really cut out for the military, too kind-hearted. The arrival of Major McLaren was another thing altogether. He was about the same age as George and had, I believe, been a boy soldier with George in their early days.

He <u>was</u> army material, keen to smarten the place up etc, Fortunately, I was now in my last six months of service and on regular's money. This was the princely sum of £2-10/- per week. whereas it had been £1-10/- For the first 18 months, however it wasn't long before I was hauled up in front of McLaren and ordered to explain my frequent absence from barracks. Needless to say, it was useless to try to explain the situation to him, as far as he was concerned the military took precedence over everything and so my grand 'skive' came to an end, shame really as I wasn't allowed to finish the course; so near yet so far!

Now, about this time, I was informed by Eddie Dow that we were to receive six new Triumph 500cc side valve vertical Twins based on the civilian Triumph Twins which were of course OHV; for evaluation purposes. The RAF already had these machines, and we were chosen due to Eddie's extensive motorcycle experience over many years. They gave little trouble being new of course, and I believe the army adopted them to replace the ageing girder forked machines.

I was now approaching demob quickly and with only a few weeks to go, I was informed that I would be required to attend the dreaded monthly RSM's parade on the main barrack square. This was a nightmare, and so far, I had managed to avoid it. I appealed to our own REME Warrant Officer that it was bad enough to undertake it early in one's army career when one's equipment was in reasonable condition but my battledress webbing etc was in a sorry state, and that I was sure to end up on a charge! No avail, he and I were never the best of friends. Something desperate had to be done and again I was saved by inspiration.

I would go on sick parade, only it would be to the dentist; now my teeth have never been of the best and I knew that if any dentist probed about in my mouth, he was bound to find a hole requiring filling. so, when the medical officer asked me what was wrong, I said I'd had no sleep last night due to terrible toothache. He, of course, asked me where it was, and I had to be very vague moving my fingers all around not quite sure exactly! Sure, enough he found a hole and filled it, job done. Upon return the WO pounced on me, threatening all kinds of punishment but upon checking with the MO my story was substantiated! *Fait accompli!* My relations with him were very strained after this but I only had a very short while to go.

During this period, I had acquired a three-wheeler Morgan of about 1927 vintage with two speeds driven by twin rear chains. It was discovered in a bramble patch by its owner after about 20 minutes search, Whilst Richard Symes and I waited. I gave him £10 for it and limped it home in low gear. the other chain was off, did it up a bit and later sold it to Peter Gill. My final vehicle was a 1934 Austin Seven which was my first true car which I used to visit Mary Lehee, a girl from Ash that I had met in a pub whilst drinking scrumpy cider. I took her home on the Scott, it was a wobbly ride, long before the breathalyser!

One day, as I was wandering about the outskirts of the motor transport compound, I came upon a rubbish dump upon which resided an unusual motorcycle. Further investigation revealed it was a Matchless silver arrow, 350cc narrow angle 26-degree vee twin. When I mentioned it to Richard, he expressed a desire to see it, so a visit was arranged whereupon he fell in love with it and it was left to me to find the owner if there was one! Further investigation at the company offices and a Staff Sergeant came forward claiming ownership. He wanted 30/- for it and produced a logbook but whether or not it was his name in it we didn't ask. After some work Richard got it on the road and used it for some time (I have a photograph of it after restoration).

Well, finally demob arrived, and I made sure that I handed over as much kit as I could, making it as difficult as possible for the TA authorities, this it will be seen was to pay off later!

Meanwhile, an orderly from the CO's office came over to see me and asked what disciplinary actions I had'?

"None," I said, he refused to believe it with my record, and it took me some time to convince him.

So, I left with the good wishes of the lads and the not so good wishes of the management and my army book!!! Which I still have, saying what a good soldier I had been etc and recommending me to any employer!

The TA unit I was to report to was HQ661. Air Op SQN.TA.RAF Station Kenley Surrey and I duly arrived and checked in with them before returning home.

It would appear that I was one of the two REME vehicle mechanic personnel and our job was to service the trucks etc for the Royal Artillery (RA) unit that was there to defend the airfield etc against mainly air attack. They seemed quite keen at playing soldiers at weekends and seemed to think that I would be joining them, which I quickly discounted, explaining that I was about to start an Aeronautical Diploma course, for three years at the College of Aeronautical and Automobile Engineering in Chelsea, and that my studies would be all absorbing. Idiots! However, to finish with my involvement in the military, I was unable to avoid the fortnight training exercise once a year, compulsory time off that the employer had to give! And at that time for three years, cheeky buggers! Here we were ten years after the war for heaven's sake! Mind you just about this time, the six days war in the Middle East broke out when Nasser closed the Suez Canal; invaded east, only to be thrashed by the Israeli's. It resulted in petrol rationing and the possible call up of the TA! Panic! I'd only just got out,

Fortunately the Yanks intervened and the whole thing fizzled out, but petrol rationing went on for some months.

It was at this time that Richard purchased an Austin Seven with the idea of turning it into a special, so with his customary enthusiasm, he immediately stripped it down to the chassis, lowered the steering and suspension, bought some cam followers, and started to tune the engine. then it all ground to a halt, he became engaged to Anne Harding whom he had met on his return to Trojans to finish his apprenticeship! Fatal. I made him an offer of £25 I think, and preceded to finish the job, using it extensively later to go to college at Redhill Aerodrome – ref photos.

Meanwhile, I reported as directed at the appointed time, to be told that we were off to Germany, a place called Buckeburg near Mindon, Hanover and that we were to fly there. A rumpus started when we lined up on parade for initial inspection.

"Where was my belt, gaiters and boots, etc?" I had arrived wearing only battle dress, shirt, cap, brown shoes and civilian suitcase with my unmentionables etc inside it. I explained as patiently as I could if it had all been handed in at Aldershot when I was demobbed.

"But", they said, "you should not have handed it all in."

"Well I don't know anything about that," I said, "I just did as I was told!"

"In that case," they said, "we will have to kit you out here".

Oh God, I thought. Again, just as all seems lost, the fates intervened, a message arrived to say that the flights had been cancelled and we were now all going by boat! And we had to get to Kenley station, train to London, thence to Harwich, for a night crossing to the Hook of Holland, immediately, no time to be lost.

An amusing incident on Paddington Station springs to mind. we were all, the TA members that is, laying about on our luggage on the platform, when two 'red caps', military policemen (MPs) arrived, expecting us all to leap up and stand to attention.

"Where do you lot think you're going, on holiday?"

"Yes," some bright spark said, "we are TA and we are going for two weeks in Germany."

They gave up at that and after some further inane banter, wandered off to persecute some other poor sods. Arriving at Harwich, we embarked on this troop ship and were expected to sleep in hammocks. Well, they are alright really; you have to sleep on your back and not move too much.

The two weeks spent in Germany were alright really, my REME mate and I didn't do a lot, most of our time was spent playing poker with the cooks during their time off shifts and anyone else who was available. We had no real workshop to do anything constructive in, and spent all the time checking oil levels, greasing and generally maintaining the various vehicles, prior to some exercise that the RA was going on. The final passing out parade was my final clash with the military mind, and it happened like this. Some high-ranking officer was leading the inspection, with various 'officials' following in his wake. When he arrived in front of me, with my part civilian gear on, he looked down his nose, turns to the Sergeant Major alongside him, and said, "take that man's name".

The SM duly started to get out his notebook, however the officer following nudged him and shook his head! Which just goes to show that there is intelligence somewhere; I bet he was a National Service Officer!

I returned to continue my studies, heard no more from the military as I think the TA was being phased out, in fact National Service finished about 1962. After coming out of the army in September 1955 and starting College in January 1956 as previously mentioned, I thought, naively as it turned out, that I would have three months rest! My mother had other ideas, "if you think you're going to lie about here for three months, you've got another thing coming, get yourself down to the Labour exchange (job centre now) and find a job".

The result was that I went to work for 'Kiddiecraft' Purley, makers of wooden children's toys, driving an electric van between their depots and later graduating to a Dormobile and removal van, operating to the London docks, and Arundel, etc.

R.E.M.E. 1953 to 1955.

News Events

Eddie Dow

Well-known Gold Star expert and exponent Eddie Dow, credited as the man who created the Rocket Gold Star, passed away on March 17.

An all-round motorcycle sportsmen – successful in trials initially, and road racing too – his biggest achievement was winning the 1955 Clubman's Senior TT on a DBD34.

Following his TT win, he opened a motorcycle dealership, switching to cars in the late 1960s, before selling the business in 1991. He filled his retirement with skiing, gardening and as a motorsport enthusiast. He is survived by this wife Diane, daughter Franca and three grandchildren.

Eddie Dow on his way to winning the 1955 Senior Clubman's TT.

Richard Symes - Silver Arrow Matchless and Mike Lane's Velocette circa 1955.

Pat Curry and me with my Austin 7 in 1955 after demob.

Chapter Five

My Time at College 1956 to 1958 – Caae

The College of Aeronautical and Automobile Engineering (CAAE) also catered for agricultural students and was situated primarily in Sydney Street, Chelsea, London, with further premises in the form of a large hanger plus offices at Redhill Aerodrome, Surrey.

Prior to this, circa 1920/1930s, the college at Chelsea was also situated at Brooklands near Weybridge, Surrey and was the first of its kind in the world! It was known as the University of the Air and had an international character with students from the British Empire and the Continent as well. The diploma granted to successful student graduates at the close of two and a half to three years study will be a degree in aerial engineering and all that pertains to civil and commercial aviation.

These latter statements are taken from *Flight over Everest*. The book about an expedition in 1933 financed by Lady Houston of which I have a copy. Very interesting. The college serviced the two Westland PV3 aircraft fitted with Bristol Pegasus engines for the 1933 first flight over Mount Everest, which was such a success. During those years, the college also supported many other aerial events like the Kings Cup air race etc, providing tuned up aircraft for wealthy aviators.

I left Kiddiecraft at Christmas 1955 and went to Chelsea in the first week in January 1956, into the aero light engine section under a Mr Ames. I was put here because unlike most new students, I was older and had had previous experience both at Trojans and the REME. My long-suffering father gave me an allowance of 30/- (£1.50) per week to cover everything except food and board, so I was still on 30/- and would remain so until the end of 1958, in other words I had been on this amount for eight years!

About this time, I purchased from an antique shop in Surbiton, a Norton model 19 of 600cc capacity for I think £10.00. This was collected the following day when we all went back in my Austin Seven car. It was decided for some reason that John Symes would ride it back and it was impressed on him that as it was of unknown quality, he should take it easy; all a wasted effort, he disappeared at high speed and Richard and myself saw nothing of him until we got home and saw the smoking machine.

"Christ, it really goes," he said, "I saw 80mph on the clock!"

Idiot!

I checked it all over and proceeded to use it for the time being, eventually fitting a Rudge type spring sidecar to it, to help keep it upright on icy roads (ref 1962/1963 winter), later on! Whilst testing it out, on one occasion, with Richard a passenger on the basic chassis; there was no body, just a plank or two for his support, we started to overtake an articulated lorry on a narrow lane in Woodmansterne, when a car appeared approaching on the other side of the road. There was no room for manoeuvre, so I shouted 'duck' which he did and I put him and the entire sidecar under the trailer and between the front and rear wheels, keeping it there for about 100 yards or so, juggling the speed accordingly, until we could continue overtaking, I don't know if the lorry driver even noticed, Symes wasn't too pleased.

Anyway, worst was to come, I decided to go for a test run up to Banstead and back, past the Oaks Park, through some sweeping bends etc, as I slowed up, suddenly a police car swept by and waved me down, they had kept their siren quiet until level with me, crafty so and so. They said I was doing 77mph in a 30mph built up area and regarded it as dangerous driving! Driving without due care and attention would have been a more reasonable charge but to no avail! I got a £10 fine and six months driving ban. What a way to start my college days.

So, I travelled each day from Wallington station to Victoria, then took a number 11 bus to the Chelsea Town Hall, for six months. During my time here, I completed courses in Light Engines which concentrated on De Havilland Gypsy series and other Rolls Royce Continental, flat 6 type arrangements which are now almost universal for the smaller aircraft due to their almost perfect balance, also aircraft electrical systems, welding, machine shops; including lathe work, milling, shaping etc, drawing office, Dynomometer testing with the automotive students and the metallic testing of metals under a Mr Beaumont. He was a very clever chap and I believe had perfected some techniques allied to proof stress etc, but as a lecturer he left a bit to be desired. Too impatient with stupidity I think, but his assistant, name I can't remember, was just the opposite and explained in detail later!

Beaumont also wrote and was involved in many books on aviation subjects, some of which I have, in conjunction with several of the masters like Cappleman, Redman and our college principle at that time, a Dr G.D. Duguid. Later, I heard that Beaumont had committed suicide, a great shame but he always seemed to be in a hurry!

Around this time, a Turkish student gave me a big heap of scented tobacco, which I smoked in my grandfather's old pipes, it was horrible, so everybody said, but I persevered; it was free! We used to go to the cafe next door to the college for the mid-morning breaks and although they had a jukebox, they also had a Mellodrome, which is like an old fashion musical box with the brass drum rotating slowly and the raised tags 'twanging' the musical leaves. You change the tune by changing the brass drum. I loved it and insisted on putting a penny in the slot most times I went in there.

As a result of my driving ban for six months, I was obliged to seek alternative transport, and was therefore riding in Chris Hayward's MG sports car coming back from the Stephen Langton pub at Friday Street, when on approaching the Swan and Sugar Loaf pub in South Croydon, we were overtaken by an over exuberant Geoff Gunter in his father's Morris car.

The road was wet, he lost control, swerving across the road into an electrical supply box, ripping it out of the ground, leaving a two-foot hole in the pavement with sparking wires protruding.

In the melee that followed; the rest of us lads arrived, and huge crowd gathered, arguments ensued as to who was to blame, a policeman appeared and started to take notes, David Oatley appeared out of the Swan and started to threaten various bystanders who were claiming dangerous driving, and John Symes fell into the hole, dancing about amid sparks, as nobody would lend him a hand to get out!! The final result which went on for several days, was that the tram system through the Croydon area was disrupted as this electrical box was part of their control system: chaos! I don't know if Geoff was prosecuted or whatever, but we all eventually retired to the expresso coffee bar, where we were originally going, to sober up!

Now at this time my girlfriend was Pat Curry, one of the Foresters Drive seat mob! (Ref list.) She was an only child and when her mother died, she and her father moved from Wallington to a flat in Balham. He was a licenced aircraft engineer and gave me some books etc which I still have.

I used to visit her, as Balham was on my way to Chelsea by road, later on when I got my road licence back again! She went to Durham University and eventually I lost track of her, shame really, I would love to know what became of them, I always thought my mother would have liked me to marry her, which leads me on to further revelations of a romantic nature.

I had previously been going out with 'Anita Chaffee' another of the 'seat mob' and one day upon visiting her at Holmbury St Mary near Dorking, Surrey, her mother cornered me in the garden and asked me whether or not I was serious about Anita! We had been going out on and off since 1950! Six years. The reason for the question, I later discovered, was Anita's new boyfriend Roy Karn. He was a motorcyclist and a nice chap but had somewhat upset Mrs C! Well, I ummed and ahhed, not knowing quite how to answer, eventually my wits returned, and I explained that I was about to start a new career at college and that my present girlfriend was Pat Curry. Say no more, she vanished at high speed. I later discovered another reason for the question and that was that her elder daughter Janet had announced her intention of marrying Arthur and emigrating to Perth in Australia!

During one of my visits to their house at Holbury St Mary, on a winter's night, I was crossing Tamworth common on my first Scott motorcycle when I came up behind the car travelling at about 40 mph, I flashed my lights and went to overtake but as I got alongside he suddenly turned right! I bounced off the side when next moment I was rushing at 40 mph through dense undergrowth praying that I would hit nothing solid, like a tree trunk. When I finally stopped, I was on my own in the 'darkling wood' as they say. Eventually sounds of rescue were heard and the car owner and passenger arrived and between us we dragged the Scott back onto the road. He accepted the blame as he hadn't indicated a right turn up a small lane to his house, his car was unmarked, but my machine had lost a footrest, had a bent handlebar lever and other small subsidiary scratches etc, Luckily the radiator had not been punctured and I was able to continue.

Later when I contacted him with a bill for the damage, which I had repaired myself, he refused to pay up and tried to blame me! Typical, I ignored him and forgot about it. This was the era of London Smog's before the Clean Air Act was passed and I remember on several occasions in late autumn and winter, in the dark evenings going home from Chelsea across the Albert bridge, through Battersea across Clapham Common and down through Balham, Tooting and Mitcham, creeping along in bottom gear on the bike with my goggles up trying to see the curb, it was that thick! With a long trail of motorists following behind me as they could virtually see nothing at all. When I got home my face was filthy black with grime and my eyes stung from the acidity of it!

In the summer of 1956, Richard and I decided to go on a tour of Wales camping holiday in my Austin Seven car, but the weather was bad and we spent most nights sleeping in the car. We toured Mid and South Wales visiting mainly old castles and historical relics as Symes was very interested in archaeology, I preferred astronomy!

In 1957, we again went on holiday, this time to Scotland, Cape Wrath was our objective and we got as far as Durness on the north coast. I had purchased a 1939 Morgan, four-wheeler with a Coventry climax, overhead inlet and side exhaust valve engine, the money £200, having been loaned to me by my father who considered the succession of 'heap's' – mainly my Austin Seven, special I built, more of this later – to be a disgrace and an eyesore when standing on his drive at Ingleby Way in full view of the neighbours!

Our intention was to go up the A1, known at that time as the 'Great North Road', remember no motorways existed until the M1 in 1961 and then at Scotch Corner we would branch off on the A66, crossing the country to the west side and then proceed to Hamilton in Glasgow, where Richard's sister, Betty lived with her husband, Ken. We had two weeks holiday and needed all of it to complete the trip, it took us almost a week to get to Durness and almost a week back! To explain, we got to Baildon on the first day, my idea as I wanted to see it since I hadn't been back since the war. We camped the first night in the farmers field, you could do things like that in those days and his wife sold/gave us eggs milk, bread, etc, very kind, didn't even think of charging us for the campsite! We slogged on up through Cumberland and southern Scotland and after a couple of camp nights, arrived at Betty's. We had a short stay here and then tackled the real task – Scottish Highlands! We went up the west side, crossed various ferries, where there are bridges like the Stromeferry and Ballachulish ferry, now bridges.

The roads were tolerable as far as Ullapool, but from there they became single track with passing places and grass growing in the middle, as it was wet a large percentage of the time, the large aluminium oil sump dragging through this grass kept the oil below its usual temperature!

At Loch Assynt we camped and were woken during the night by the tent taking off in the wind and rain, we eventually held it down with rocks! Another ferry at Ullapool, the Kylesku, now a bridge and we finally arrived at Durness. The weather by now was worsening and our camping gear soaked, so we decided on a bed and breakfast for just two nights, but the next day the weather was so bad we took off, eastwards around Loch Eriball towards Tongue and thence southwards through Altnaharra and Lairg. The sports car hood leaked at the joints and water was running down the dashboard onto our knees! We wondered how long it would be before something shorted out and the engine stopped. After one more miserable night we arrived back in Glasgow at Betty's, completely soaked and fed up. During this time, we visited Edinburgh as Richard wished to see the Castle and I believe the Edinburgh tattoo – a military pageant – was on or about to start. One amusing incident occurred; we were standing on the pavement, me with my army dispatch riders leggings and battle dress on, remember I had only just come out of the Army in September 1955, when a large staff car swung around the corner with Field Marshall Montgomery sitting in the back seat on my side with the window open.

Well, it took me completely by surprise and I said to Richard in a loud voice "look, it's f***ing old Montgomery!"

He glared at me, said something to the driver and the car slowed down but it was too late, we crossed the road and disappeared into the crowd. I often wonder what would have happened had they caught me, technically I was demobbed but still in the Territorial Army! Who knows? I never liked that man anyway. We stayed a couple of days here to get sorted out and then set off south for the Lake District as neither of us had ever been there.

The weather by now had improved and we had some sunshine, the only real thing I remember about the lakes, was going up the Honister Pass and worrying that we would not make it as the car with its relatively high bottom gear and the state of the road/track, meant I had to slip the clutch frequently! After, this we returned via Baildon and stayed overnight with the Bottomley's, he was the man who stopped to help us on Ilkley Moor years before; they had lived opposite us in Ferncliffe Drive but had now moved to a larger home in Greencliffe Drive. They gave us a bed and breakfast.

By now, I had been transferred to Redhill Aerodrome for the more practical work on actual aircraft. We had several Tiger Moths, Austers, etc and a large twin-engine Percival Q6.

I started here in the woodworking section under Mr Redman, where we made from scratch a section of a wing; following this, I transferred to the metalwork section under Mr Cappleman and we again made a section of wing, this time using aluminium sheet, rivets etc, instead of the wood Sitka Spruce, glue, fabric etc, before we were let loose on the real aircraft! While all this was going on, we had lectures on the theory of it all and were expected to compile notes into a large folder that would be part of the final exams i.e. coursework, it's called today, I believe.

The larger engines such as Bristol Hercules, a 14-cylinder radial Merlin, and some gas turbines, plus De Havilland hydromatic variable pitch propellers were studied, this was very important to me later as you will see! Comprehensive classes were given on aircraft instruments, theory of flight and aircraft construction Etc.

I started late 1956 here at Redhill using the newly completed Austin Seven special during this time of the Suez crisis and petrol rationing and in order to make it go further for my five day a week journey, which was further than my visits to Chelsea previously, I ran it on a mixture of petrol and paraffin!

I used either Aladdin pink or Esso blue in the proportions of one petrol to two of paraffin, the former being slightly the better of the two in general performance. No initial priming was needed with this mixture but if total paraffin was tried, it required a float bowl full of neat petrol to get it started, which could be easily achieved by using a model aircraft type fuel filler directly inserted into the carburettor float bowl; but it did smoke heavily for a while until it fully warmed up.

This was brought home to me on one occasion when coming home from the aerodrome, a speed cop on a motorcycle came up behind me and as I had only just started off, I feared he would smell the exhaust but either he didn't, or he was a decent chap! Taking pity on a poor student.

On another occasion, coming home along Smithenbottom Lane in Coulsden, late one afternoon, the offside front mudguard of the motorcycle type, dropped onto the front wheel, wrenching the steering wheel from my hands. Luckily, nothing was coming in the opposite direction as the vehicle veered off through a front garden hedge, ripping a large section of it, turning the car on its side and slewing onto the front lawn/garden of posh large house! I scrambled out and a man came running up the road, nobody else was around, just shows how little traffic there was in those days. We righted the car and pushed it out of the garden and onto the road. The front axle was bent back on the offside; radiator was losing or had lost all of its water, it started up all right on the handle, so I told the man to keep quiet, nobody came out of the house, so I drove it home with slightly weird steering and went back the next day on the motorbike, noticing the large hole in the hedge and torn up flowers littering the garden but I was too much of a coward to face the music.

These were happy days – more so than Chelsea because I was out in the country and the run down there was also ideal for motorcycling, down through to Coulsden onto the A23 to Merstham and then the back lanes to South Nutfield where John Collett once lived. I used to use it as a racecourse and had one or two near ones, nearly colliding with a woman pushing a pram and crashing in front of a herd of cows that were blocking the road, I had to lay it down or hit the cows which could have been a costly business! Also, during the lunch hour and after classes finished, I would practise road racing around the perimeter track on the Norton, this was great from my point of view but the authorities were not so keen, the aerodrome was after all still operational with some regular services to places like the Isle of Wight using De Havilland Rapides, Herons and Doves were used, also private flying was taking place.

It eventually came to an end when I overshot a corner, veered onto the grass and collided with the shaft, hidden in the grass of something like a combine harvester that was being used for instruction by the agricultural section of the college. Unfortunately, somebody I think in the control tower saw it phoned up the college complained officially, so that was that.

After considerable work on the damaged Austin Seven special, I repaired the machine and started using it again and on this occasion was coming down Carlton Avenue where a group of the lads were standing, I decided to show its versatility off and moved over into the passenger seat, I could now continue control by using the hand throttle lever mounted in the centre of the steering wheel and I decided to do a U-turn at the T-junction with Ingleby Way and go back the way I had just come. Unfortunately, as I turned the steering wheel it took the throttle lever with it and the engine opened up to full bore, I couldn't get the clutch out for obvious reasons and the whole contraption mounted the pavement and went through Mr Parrish's fence with a splintering crash and we ended up in his back garden where he was digging his flowerbeds.

He was pretty decent about it and said if I would repair it, he would say no more about it! So, what to do? It had taken a whole six-foot length panel out but fortunately not broken the upright supports And I realised that these were standard on almost all of the fences in the road. I went to the bottom of our garden and removed an identical panel and fixed it back in place in Parrish's fence, job done! A day or so later just as we were about to have our evening meal, my father burst in and said "Vi", he always called her 'Vi', short for Violet; "somebody has stolen our garden fence", so we all trooped out to have a look. I don't know whether Phyllis or mum knew but if so, nothing was ever said.

We had a freak storm one afternoon about 3:30pm and lightning struck the stores, which contained many metal items on shelves around the room. Although I didn't actually see it, I heard the crash and was there only moments later; the storeman was considerably shaken and apparently a ball of lightning was seen jumping from one metallic item to the next for some seconds before disappearing in a flash! This poor devil was trapped in there with it and had to have a strong cup of tea to restore his nerves!

One day, a notice went up on the board inviting all students to a dance at the Redhill hospital, to which a large group of us attended one evening. As we entered the grounds of the hospital there was a general cry of 'men' and rows of nurses faces appeared at the windows and we were cheered all the way to the entrance door! I met a nurse here called Breda Dalton, she was Irish and came from Thurles in County Tipperary. I took her out in my Morgan 4x4 several times but was unable to spend much on her as I only had my 30/- per week! Quite embarrassing, one meal or trip to the cinema and that was it. We eventually fell out over my intended participation in vintage motorcycle racing; She was opposed to all the motorcyclists, because she said she had seen so many casualties.

This was a time of wild parties especially those held by Johnny Kharaget, Pierre Jobez and Turpin who all shared a flat in Purley.

Turpin was a character, slightly older than the rest of us, he lived at Chichester called 'Chi' on the coast and on one occasion, invited Evans and me to visit him, where he lived with his mother. We had a whale of a time with him and his mates and on one occasion, when driving back from a wild party, Evans was violently sick, hanging his head out of the car window while Turpin continued to drive at high speed! On another occasion, Turpin said he knew of a pub near Beaconsfield that brewed its own beer, so we set off and indulged in this 'purple liquid' but I have no idea how we got back or where the pub was or is! Fred said he had been there also, it's not too far from Dunstable but that he also does not know where it is! And has never found it since – a bit like Brigadoon! Only appearing every one-hundred years!

I also purchased for 30/- an 'Areonica' JAP 1800cc flat twin engine to put in a three-wheeler Morgan but I'm afraid never got time to do it, my ideas were always running ahead of me! My thoughts were now turning in the direction of vintage motorcycle racing with John Lane now competing in some vintage events on his CSI Norton. Therefore, I decided to sell the Morgan 4x4 which always worried me as I had to keep it in good order to get my money back and pay off fathers' debt. A chap from the west country bought it straight away, no problem, it was a lovely thing to look at in British racing green, Chrome radiator and headlamps etc.

With the remaining monies, I purchased an Austin Eight van which had belonged to a fishmonger! My mother was horrified once more and said she would not allow it to stand in the drive on view to the neighbours. So, I had about a gallon of army green paint, pinched from Aldershot, I gobbed it all over, painting out the fishmonger's name, etc, until my mother considered it okay.

With this vehicle, I did two trips in 1958, first was with Bob Durrant to North Wales, where we camped out in a farmer's field and set out to climb Snowden. Unfortunately for us, the weather deteriorated, and it began to snow. It was Easter time – typical!

We had taken a direct route up a steep slope and had to cut footholds using my parents coal hammer, I had borrowed! Eventually, we reached the railway track, turned left and followed this to the summit where we saw nothing except blinding snow. We returned down the railway track avoiding the steep slope we had come up and returned wiser men! Whilst we were doing this, Richard and John Lane were partaking in the Land's End trial on their two speed Scott and CSI Norton, respectively.

The second effort was in the summer of 1958, when Kharraget, John Lane and myself decided to re-trace the previous year's trip to Durness that Richard and I had done. This vehicle was more suitable, having more ground clearance and would sleep three of us if necessary! Although, we preferred the tent with one of us in the van.

We set off, me at the wheel of course but the idea was; once we got off the beaten track into Scotland, John would have a go at driving but as it turned out it was a bad idea. The roads were totally unsuitable for a new driver and I was too impatient to be a good teacher so after a few stalled attempts, crunched gears and trips off the road, I took over again. Our first real stop was at Hawick in Southern Scotland, where we went to see the Jimmy Guthrie memorial to a great British pre-war motorcycle rider killed at German Grand Prix in 1938. This was at the request of John who at that time attended every TT each year!

After a day or two, we got Loch Lomond and camped by the lakeside, of course it rained, and we discovered that my tent leaked! We had my petrol army stove for cooking on and a blow lamp for additional heat, should it be required, and I succeeded in keeping the drips to a minimum for a while by playing the blow lamp over the inside of the roof! The other two took off into the van complaining of lunacy on my part but it was my tent anyway.

We now headed for Fort William as John and I wanted to climb Ben Nevis at 4,406 feet, highest mountain in Britain. We set up camp in Glen Nevis, on the southside of the mountain and tried, unsuccessfully I might add, to light a fire on a stream; we wondered why it wouldn't light! The following day, we set off at 10am up the peasant track, after climbing for about three hours, Kharaget, who was never very keen, decided to quit. We pressed on into the mist and after what seemed like ages, actually another two hours we reached the top, knackered! We returned at about 8:00pm, it had taken us all day, still we had some consolation we had climbed it before Wainwright who made all these mountains popular!

As we got more into the Highlands, the scenery got more rugged and wilder and we sighted a very large bird sitting on a rock about 200 yards away; almost at the same instant, the rear doors of the van were flung open, and Kharaget leapt out, blazing away at it with the 12-bore shotgun! I might point out here that John and I had not been very happy when we discovered that he intended to bring his girlfriend's father's gun with him! The trouble was he was an Anglo Indian from Bombay, obviously wealthy parents, who thought that England was like India and he had the right to do what he pleased. Don't get me wrong he was a nice chap, popular with his fellow students and very popular with the ladies!

The bird in question, obviously not hit, took off with a contemptuous flap of wings and soared away over the valley below, I am sure that it was an eagle, the first one I had ever seen and if it hadn't been for that idiot Kharaget, we could have observed it at leisure through the binoculars.

On another occasion, Kharaget shot a duck, we had stopped next to a loch and the bird fell into the water. As John was the only swimmer, we talked him into going in for it, it was a mistake and could have had serious repercussions. He started off and the water was stone cold, but he kept going, egged on by us two idiots, but was making no progress towards closing the gap on the bird. Eventually, it became obvious to us that it wasn't dead and was paddling away from John, he of course, could not see this when we began to shout to him to come in. He was uncertain but luckily, at that moment the bird took off and John struggled back towards us. He made slow progress, he was some way out and we began to realise that he could be overcome by exhaustion or cramp! His teeth were chattering when we eventually wrapped him in a towel, lit a fire and camped there for the night.

We pressed on northwards, past Loch Assynt, where Richard and I lost the tent the year before but as previously the weather got worse, wet and windy. We were reluctant to put up the tent but there seemed to be no accommodation like a Bed and Breakfast. We were miles from anywhere, somewhere near the Mountain Arkle, if I remember properly. Suddenly, we came upon a large barn which was open, so we drove the van and everything inside and shut the doors. We stayed here a couple of days, dried everything out, checked the vehicle etc and during this whole time we saw nobody. We left it as we found it, in good order and pressed on to Durness.

This time the weather was good, and we camped in another old barn on the outskirts of a village. Again, nobody bothered us, and we spent our time exploring the Smoo Caves, walking and fishing! John, I think, had bought a rod and line and our first attempts in the sea were to produce many small sprat-like fish, which we threw back considering them too small. But after some time, another angler arrived, and he caught the same thing but put them in a bucket. Upon investigation, it turned out that these were considered a delicacy, so we followed his advice. We weren't sure what to do with them, so we fried them and had them on toast!

We also had a cooking problem with a Fray Bentos meat pudding, which said on the tin 'cook in an oven'! We had no oven, so we fried it, chopped it up into three and ate it! In the afternoons, we used to walk back into Durness and have tea and cakes in a small cafe cum tearoom, which was really somebody's front room, all very informal, we told them where we were camping, and they said not to worry stay as long as you like!

From here, we went east to Tongue, south to Altnaharra and Lairg, Following the previous year's route but then we went east again at Inverness to Aviemore and into the Cairngorm mountains. It was here that Kharaget got his largest 'bag', he rushed off into the bushes one day and we heard loud reports, whereupon he returned with an armload of dead grouse. Apparently, they were sitting on the nest when he arrived, and he just blasted them at point blank range! Very unsporting and dangerous because it was before the glorious Twelfth of August! So, we hightailed it out of there, Post haste.

Later that day, when we camped for the night by a river, we decided to try and eat one of them, but none of us had any idea of preparation so we pulled off the feathers, chopped off the head and tail portion and held it in the river to clean out the insides by letting the water flow in through the head end and out of the tail. Then we boiled it. It tasted alright to us famished campers!

Now, we started for home, but the van was giving us trouble. It turned out later, upon inspection, that the exhaust valves were burnt out on some cylinders, they had flats on them and were D-shaped. Every morning we had to push start it so we made sure that we camped on a hill so we could run it downhill to start. However, when we got into Yorkshire it was getting so gutless that it would not climb some of the hills in bottom gear!

This meant that John and Kharaget had to sit in the back and open the rear doors, leap out and push, with me, sometimes even slipping the clutch in bottom gear. Eventually, we got onto the A1 again, a reasonably flat road, pulled up at Jacks Hill Cafe near to Baldock for breakfast. We had been driving through the night, not wanting to stop the engine for obvious reasons. This time I jammed a piece of wood between the throttle pedal and dashboard so that the engine was on a 'fast' tick-over, we had a quick breakfast. It was so bad, I remember struggling to get up Foresters Drive, which was only a mild slope in bottom gear with only myself in the vehicle at about 10 mph!

Finally, the following day my mother cooked the remaining grouse, about six, if I remember correctly, they were about a week old so just right in some people's eyes. John and Kharaget joined us and they were bloody good!

I have digressed somewhat from the college days so to get back on track; while still at Redhill aerodrome we had built up from scratch a Tiger Moth and this machine was now ready for its C of A. Certificate of Airworthiness. Whilst trying to start it up, by swinging the propeller, it kicked back and the opposing blades caught the Indian student, who was providing the muscle, across the arm. There was a resounding crack, heard all across the hanger and his arm was broken. Apparently, Eastern bones are more brittle than Western ones, due somebody said, to the different diet. A test pilot came down from the ministry and I was chosen to accompany him, taking notes and readings in the front cockpit whilst he threw the machine all over the sky on its first ever flight! He let me take the controls and make gentle turns etc before landing it. It was a very interesting experience, especially as I had helped to assemble it! Finally, my time at Redhill and Chelsea came to an end and I was affiliated in Morton Air Services at Croydon airport which was still operational at the time.

The idea of this was to give the student, what is now known as 'work experience', and therefore by arrangement with various aircraft manufacturers, servicing facilities, etc we spent the last three months of our college time here. By 'we', I meant Johnny, Kharaget, Pierre Jobez and myself. At that time, Morton's serviced several private aircraft, in addition to their own fleet and one of these belonged to Douglas Bader, the well-known, legless Battle of Britain pilot. Unfortunately, he treated the mechanics with disdain and was generally disliked.

Even more unfortunate was that at our graduation ceremony, he was to present the diplomas and I, of course, therefore refused to go! Mine arrived several weeks later by post. In retrospect I wish I had gone for my parent's sake, I feel that they were a trifle disappointed, that's typical of my stupidity!

Anita on Mike Lane's Velocette with sister Janet's
Royal Enfield in background circa 1950s.

My Austin T Special taken at Redhill aerodrome when at C.A.A.E. circa 1957.

NOISY MOTOR-CYCLES ANNOY RESIDENTS

RESIDENTS of Ingleby-way, Wallington, are often annoyed by young men riding noisy motor-cycles at night, said Police Inspector A. Ewart at Wallington Court last week. Robert Cherry, of Chipstead-way, Banstead, and Robert Collett, of Ingleby-way, were summoned for riding motor-assisted pedal cycles with inefficient silencers and without lights.

Collett was also summoned for riding his machine without efficient brakes. He was fined a total of 25s. Cherry was fined a total of 15s.

P.c. D. Harrison said that after receiving complaints, he went to Ingleby-way. He could hear defendants' machines from some distance away, because of the excessive noise they were making. When he saw them he noticed they had no lights.

GAPING HOLES

He found that Cherry's machine had a piece of copper tubing fixed to it instead of the normal exhaust pipe. There were gaping holes in the tubing and it had little or no effect as a silencer.

Told he would be reported, Cherry said he had only received one complaint, and that had been three days before.

Collett's machine was also fitted with copper tubing as a silencer. It had no front brake and the rear brake was inefficient.

Inspector Ewart told the court: "There have been serious, persistent and well justified complaints from residents in this area about lads riding without lights and making excessive noise. Local residents are seriously perturbed."

'TRIED OUT' NEW M/CYCLE —AT 75 m.p.h.

Stopped by police after riding his motor-cycle at 77 m.p.h. along Croydon-lane, Banstead, Robert George Collett (aged 22), of Ingleby-way, Wallington, told them: 'I only bought the bike last Saturday and came along here to try her out. I didn't realise that there was a 30 m.p.h. limit here."

Police stated that Collett, a student at the Royal Aeronautical College, overtook three vehicles while travelling at 77 m.p.h. and passed several children playing in South-drive without slackening speed. He also "took several very dangerous bends" at 60-65 m.p.h.

"We take a serious view of this offence. There have been some very bad accidents along that lane, you know," chairman of the bench, Mr. E. Simmons, told him.

Collett was fined £10 with 2 guineas costs and had his licence suspended for six months.

1956

1957 at Colin Shield's wedding.

Collecting wood in Glen Nevis - 1958.

John Lane - Scotland - 1958.

Kharaget and me at the Forth Bridge - 1958.

Witchford Sprint circa 1958. Me on model 18 Norton 500cc 1929 vintage.

Chapter Six

My First Real Job – De Havilland Propellers – 1958 to 1963

For a month or two, I worked at a local garage as a vehicle mechanic, on the strength of my army qualifications and experience, whilst I looked round for a job in the aircraft industry, commensurate with my qualifications. Eventually, I came upon an advert in one of the national newspapers for a job as a Development Engineer for De Havilland Propellers at Stevenage. The interviews to be between 7:00–9:00pm at Conquest House, London. As I remember it, my parents accompanied me, and we went on to a show or film afterwards.

When I looked round at the various other applicants, they were all older and more experienced engineers, and I thought my chances were slim. However, at the interview with a Mr Ross Hall and Mr Mike Osmond, all went well, especially when we got onto the subject of the Hydromatic Propeller; this we had spent some time on at Redhill and I was able to describe its operation in detail. The reason I labour this point is that I consider it a turning point in my career, setting me off on a path towards a 'white collar' as opposed to 'shop floor engineering' future. Later, a letter arrived confirming the job, at an astronomical salary of £750 per annum or £15 per week!

When I arrived at De Havilland Propellers, I discovered that I was the only one to have been taken on and this was on account of my Hydromatic Propeller knowledge. I stayed in 'digs' four nights a week, most of the time with a Mrs Ross, her husband and six sons, ranging in age from 16 to infant. I used to cross London every weekend, usually on my motorcycle on a Friday evening and return early Monday morning.

I had just started vintage motorcycle racing at about this time and purchased a large, old fashioned, flat belt driven lathe and pillar drill, which I installed in my father's garage, also I was still partially involved with aeromodelling and had built a Dynajet Control line model which I hoped to fly but never did. It was not really a Dynajet, but a handmade pulse jet, which I had got from a chap at college half-finished and I completed it on the lathe etc.

Martin Fenny and I used to run it up on my dad's bench, the noise was fearsome, worse than a motorcycle with open exhaust, its cyclic frequency was 18,000 cycles per minute and on one occasion, when my mother was in Woolworths in Wallington, about a mile away as the crow flies, people in the shop were wondering what the strange noise was! My mother did not enlighten them!

Also, on one occasion, it blew back through frayed petals (which incidentally, I made out of razor blades at .005-inch-thick spring steel) And set the bench etc ablaze, quite a hairy moment as we did not have a fire extinguisher! About this time, I went as a mechanic to John Lane who was about to compete in his first road race at Mallory Park in Leicestershire, unfortunately we were unable to start the engine due to my stupidity in boring out the jet block of his carburettor, for greater, we hoped performance! However, it did enthuse me to race myself and thereafter we attended most meetings together.

I spent a week in the War Memorial Hospital in Carshalton at this time after putting an air gun slug in my right-hand index finger due again to my own stupidity! They used to keep you in for even minor operations in those days; now they throw you out, often within 24-hours after major surgery!

The first meeting after this was a quarter of a mile sprint at Witchford, where I did the distance in about 16 seconds but my bike, a model 18, 500cc Norton on petrol was not set up for this type of event and so from then on, I concentrated on road racing. The first race I entered was the August Bank Holiday meeting at Crystal Palace, south London. The circuit isn't used for racing now as it is considered too dangerous; there are no runoff areas demanded by modern circuit regulations, all of the bends were either concrete or railway sleeper encased!

It was an amazing race from the start. As the flag dropped, remember in those days it was a push start with a dead engine, somebody on the front row stumbled, fell sideways, and more or less the entire row fell over also. The rest of us, some who had their engines running, somehow wrestled our way through and the race was on! It was won by Wiffen on a Rudge, I was fourth, but John was lying second for most of the race until the last lap when his plug lead fell off! We were both in our first real race. Several races followed and then came Aintree, up in Lancashire, this I remember well, as I joined the horse racing circuit for a while after my rear brake rod stripped it's retaining thread on a bend forcing me to go straight on. Fortunately, there was a gap, followed by a ditch, which I jumped, next thing I know I am on the horse racing circuit between white railings on either side! I thought, well there is not much point in going back and losing valuable places, so I kept going for about 200 yards I suppose, when I saw another gap back onto the road circuit. I re-joined the race to finish around midfield! The lads thought it highly amusing. Castle Coombe provided some hilarity when someone stuck a Norton Megaphone, engine running, into the tent to wake me up for morning practice.

My first win came at Brands Hatch in 1961 where my mother and sister attended, where I am told my mother created a bit of a sensation by waving her brolly and shouting me on! Good for her, she was lovely! My father never came to watch me race, I think it worried him, being perhaps more sensitive than mum. The results were published on the front page of the De Havilland magazine, of which I have a print, which leads me onto my first holiday in the autumn of 1959, whilst at De Havilland's. At this time, I was running a 500cc Ariel red Hunter, my second Ariel after the infamous Square Four of yesteryear and John Lane, who was to go with me, had on the road a model 18, 500cc push rod Norton with sidecar to which was fitted a huge homemade box in which he used to transport his vintage racer to meetings. So, in order to carry my share of the camping equipment etc, I fitted a sidecar chassis; borrowed the hen house door from the chickens! Got a trunk out of the attic, roped it all on (my parents and Phyllis were away on holiday at the time) and took off for Wales with John.

1959 was a long hot summer, one that only seems to come along about every ten years or so in this country. We went in October and so good was the weather, we did not bother even to put up the tent most nights, simply slept out under the stars, marvellous! Our first incident, occurred at Pendine Sands in South Wales, which we visited first as we wanted to see the famous record-breaking sands where so many world records were set in the early part of the 20th century.

We parked our bikes on the sands and went off to explore, when to our horror, we saw that the rapidly approaching incoming tide was lapping around our bike wheels. I managed to get mine started and made it onto the promenade and returned on foot to assist John with his heavily loaded outfit. He had got the engine started but every time he let in the clutch, it sank deeper into the sand, the tide was by now lapping around the crank cases! Luckily, several holiday makers etc had turned up and between us all, we managed to get it out. Continuing around the rest of Wales, we had an interesting experience with the law (as usual) when approaching Bala. It was early evening and we had just come out of a cafe/pub after an evening meal getting on our bikes to go and look for a suitable campsite, when a young, keen copper arrived and started to question us, tax, insurance, no MOT in those days etc.

After a lot of argie-bargie, checking the bikes over, lights, brakes, tyres etc, wasting time, a Sergeant appeared, he seemed more sensible and told the constable to get lost effectively, he would deal with the matter. He explained that the constable was new! He had been recruited from Caterham Barracks; he was an Ex-Guardsman; say no more.

We continued North, eventually arriving at Bardsey Isle off the northwest of the Llyn Peninsula, where we proceeded to do some nude bathing on a remote beach in the hot sun. On our way back down the A5 we camped in a farmer's field near the mountain of Tryfan and I decided to climb it. John decided against it, so I went on my own. It was a long hard slog especially near the top as it became very rocky, resulting in a hand-to-hand scramble to achieve the summit of two large boulders, the views were magnificent, well worth the effort.

We returned and I faced an irate Father, complaining of chickens wandering about all over the garden, eating his plants etc, however I refitted the hen house door and he calmed down. Meanwhile, work continued at De Havilland's; I was working on several interesting projects including the Rotodyne helicopter, Bristol Britannia, Vikers Viscount etc, all involving the development of the engine propeller, hydraulic and mechanical systems. Around this time, 1960, the government decided after years of spending taxpayer's money, time and effort; to scrap the Blue Streak rocket programme and buy an inferior, but cheaper Blue Steel rocket, thus effectively writing us out of the space race. Disgraceful, but I'm afraid it's an all too familiar story of British government, of all parties! More of this later.

Now, it so happens, at the back of our factory with restricted access, they were building the Blue Streak missile structures and Ted Snook was, at that time, a De Havilland Apprentice working there, that's how I met him more or less. Well, when the decision was taken, we had the disheartening experience of seeing through our office windows, the lorries with partially completed Blue Streaks being taken away for scrap! Followed by the inevitable redundancies of highly skilled men.

This same year, 1960, saw the start of our Irish holidays, these were inspired by me, I think, and came about as a result of seeing *The Quiet Man* film circa 1954 with Mary Lehee, the then current girlfriend, also Irish, whilst doing my national service in Aldershot. It was a chocolate box view of Irish life, plus some Hollywood hocum, but it gave me a strong desire to visit the emerald Isle. By now, I was earning as well as my friends, so I put the idea round and five of us decided to go; they were Richard Symes and his wife Anne, he would go on a Triumph speed twin 500cc with a Jet 80 sidecar; John Lane, CSI Norton Single with the sidecar, as 1959 Wales trip, Ernie Evans, Triumph Tiger 100, 500cc Solo and myself; this time with Norton model 19, 600cc with sidecar.

Our intention was to see as much of Ireland as possible in a fortnight (our annual holiday at this time) taking in the Ulster Grand Prix at Dundrod, near Belfast and the village of Cong in Connemara, where the *Quiet Man* was filmed; therefore, we decided to go round the coast road from Roslare to Roslare.

Things did not go smoothly from the start, Ernie's bike was not ready, so John decided to stay behind and help him to finish it; therefore Richard, Anne and I set off on time to catch the night boat from Fishguard. As we were coming through Abergavenny, Richard lost it and mounted the pavement on a left-hand bend, causing slight damage which included bending the front forks unfortunately, without impairing their movement too much! We continued, caught the night boat, arriving in Roslare in the early hours of the morning, where we stopped at the side of a country road and cooked our breakfast.

Meanwhile, John and Ernie set off a day later, to catch the next nights crossing; an amusing incident occurred which could have been serious. When descending the steep hill in Fishguard that leads to the docks; on a right-hand bend, the real wheel of the bike elevated and would have continued to do so if the front end of the sidecar box had not come into contact with the road. He continued down the hill in this manner for some distance before regaining control. It could have been catastrophic if he had gone completely over the handlebars!

Meanwhile, we continued to make our way north passing through the Wicklow mountains, keeping to the coast road as much as possible, continuing to look for John and Ernie as arranged but with no success, we didn't know if they were in front or behind us; remember no mobile phones in those days! Eventually, as we approached the border between Ulster and the Republic, we came upon them at the side of the road. They had had a harrowing night in the rain and had to sleep under the ground sheet strung between their two bikes, as we had the tents with us!

We proceeded on, and watched the Ulster Grand Prix, where Ernest Degner on the lone MZ two stroke, took on the might of Honda with riders of the calibre of Redman, Hailwood, Phillis, etc, and unfortunately for us, two stroke enthusiasts, Richard and me, he was beaten into second place, I think. Afterwards, we all went into Belfast (remember this was before the trouble started in 1968) for our evening meal. I remember this clearly as for 2/6d we got a big steak and onion pie (1/-) and a pint of Guinness (1/6d) and you didn't need much else after that lot! Lovely. We continued on up the Antrim coast road north then turned west towards Donegal, down through Sligo and into Connemara and Cong, where we visited The Quiet Man cafe and were able to recognise various places seen in the film. From here we went to Limerick and in a cafe here received very off hand treatment! The reason, we later understood, was because Oliver Cromwell had 'sacked' the town during the troubles, Some three hundred years before!

We completed our circuit of Ireland by way of Killarney, where I climbed Carrantuhill the highest peak in Ireland at 3,414 feet. We all started off to climb it but John and Ernie both not very sure footed, came to grief on the Nag's Staircase and retired. Richard and Ann kept with me for some time before Ann tired and they also gave up. I scrambled up an almost sheer cliff, using both hands and feet to the summit. There was a tin box at the top besides the usual statue of Mary! And in it was a book to record one's name and address, plus comments etc, which I did. Many years later, when Jill and I climbed it in 1991, I looked for it but no sign, later at a local pub we were told that hooligans had thrown it off the top! Typical.

Finally, to round off the holiday, we arrived back in England and coming back along the A40 near Oxford, I think, in pouring rain all the way, we pulled up quickly as the traffic lights turned to red and John shouted, "here comes Evans," and the next thing I knew was a vision of a bike on its side sliding along the road with Ernie following it on his backside!

Being the only solo machine, he had lost it during the sudden breaking, however no harm was done to either man or machine and the rest of the journey preceded without incident. Meanwhile, back at De Havilland's, I had been selected with several other of the younger members of the Development Test Administration (DTA), to fly in a Hastings; four-engine propeller driven aircraft with the inboard port propeller as the test one, used mainly for de-icing trials. I have a photograph of the machine from the De Havilland magazine.

In order to qualify for this, we all had to undergo a medical examination at RAF Hendon in London. It was fairly comprehensive as I remember, particularly in sight and hearing tests. Three flights stand out as memorable, the first two were flown from Hatfield Aerodrome, where the main De Havilland factory was.

1. On the first occasion, we had reached about 13,000 feet (we went on oxygen at 10,000 feet as the aircraft was unpressurised) and we were happily spraying water on the aforementioned propeller, when there was a loud bang and the fuselage filled with smoke! We all panicked of course but after investigation, it was discovered that an air pressure line

under the floor had broken and the resulting loss of high-pressure air had blown dust all over the place!

2. The second incident was more serious, we were again spraying the propeller and all of a sudden, the pilot or co-pilot reported, over the headphones we all wore, that the controls of the aircraft we're not responding, and he was flying using the trim tabs! As there was a severe danger of complete loss of control, possibly an ever-increasing risk of a nosedive, he had sent out a 'mayday' call and we were ordered to come forward with our parachutes fitted and gather around the hole in the floor about four-foot square through which we were to jump if/when necessary! Bloody hell I thought we've had no parachute training at all, let alone any practice jumps. Unbelievable! Well, to cut a long story short, we stooged along like this for a while, slowly descending, trying to stay straight and level until at about 4,000-foot altitude, the pilot announced that full control had returned, and we made a normal landing at Hatfield amid a load of emergency vehicles etc.

    It transpired after investigation, that one of our water pipes feeding the spray bar which ran under the floor where the control cables run, had broken. The resulting water then froze at 13,000 feet, seizing the controls until we had descended into warmer air at 4,000 feet, when of course, it melted!

3. The third episode occurred at Weybridge Surrey, home of Vickers Aircraft on the old Brooklands racing circuit used before the Second World War. Rex Boyer and I were sent down there to do strain gauge stress work on the starboard propeller of the twin-engine Vickers Viscount aircraft, which had a large bundle of wires taped down across the leading edge of the inboard wing, on the starboard side.

    The aircraft was being flown by Vickers chief test pilot – Mike Lithgow, and he could tell a few scary yarns.

This time we had just taken off and reached an altitude of not more than 1,000 feet when all of a sudden, the aircraft started to vibrate heavily, and Lithgow sent out an immediate 'mayday' call. Looking out at the starboard wing showed that the entire cable ducting had come loose and was 'flapping in the breeze' so to speak. We were on intercom with the pilots and could hear all that was going on.

Lithgow now reported that he was losing lift from the starboard wing and this was evident from the list that we had developed! He reduced the revs of the engine to reduce the flapping vibration and started to feather the propeller. Immediately the list worsened and we were in severe danger of side slipping into the ground!

I heard Lithgow shouting to unfeather quickly which he did, and we continued to turn through 360 degrees for an immediate landing! Which we did again amid a load of emergency vehicles etc.

Later, in the pub down the road, Lithgow said, "it's not often a test pilot earns his keep, but I think we did today". Unfortunately, he was to lose his life a few months later when the BAC III tail stalled into the ground and before the cause of this was understood for tail engine aircraft.

The reason for our problems, apart from the obvious vibration due to the loose wires, was that this bundle flapping about on the upper side of the inner starboard wing had disrupted the vital air flow over about 1/3 of the wing resulting in a severe loss of lift! When Lithgow had tried to reduce revs and feather, a further loss had occurred causing a greater list. So, we had to limp back, accepting the vibration, hoping the aircraft would not fall apart. Goodness knows what would have happened if we had gone any distance from the airfield before it happened? You can now see why I am nervous of flying; I hear every noise and bump!

Meanwhile, we, that is Richard and Ann Symes, John Lane and myself decided to go again on holiday to Ireland in 1961. Evans, for reasons best known to himself decided not to accompany us this time. We went, with sidecar outfits, Richard's Triumph Speed Twin, John's CSI Norton and my model 19 Norton, and we again sailed from Fishguard to Roslare.

The crossing was terrible, the boat was all over the place from the word go, both Richard and Anne were sick and the smell in the lounge and anywhere below deck was appalling! John and I eventually began to feel queasy, so we upped and went on deck and slept in their scuppers in spite of the rain.

We set off north, as before in 1960 but this time we went to Dublin and saw Nelson's Column before it was pulled down when the trouble started again in 1968! While we were there, we visited Stanley Matthews shop, although we didn't see the great man himself. From here we purchased a Norton ES2 500cc single; whereupon we stripped it to pieces bringing home the engine, gearbox, clutch, wheels etc leaving only the frame and Forks! Also, we visited Glendalough in the Wicklow mountains and camped for the night. Before retiring we visited the local pub where Symes was disappointed to find they did not sell cider, exclaimed in a loud voice "what no cider, peasants!"

There was a group of men sitting in a dark corner who kept looking at us for some time after this and we thought 'trouble'. No mistaking our English accents! Eventually, one of them came over and asked if we would join them for a drink! After that everything went well, they wanted to know as much as possible about England.

Whilst here, we climbed the Sugar Loaf Mountain which lies close to the coast just east of the main Wicklow Mountains and were rewarded with great views all around and of Wales and Snowdonia etc. We spent a large part of the time racing from one area to another with the motto 'the last one there buys the drinks', so we didn't see as much scenery as we should.

It was during one of these crazy 'thrashes' that Symes modified an Irish gentleman's push bike, it happened; thus, we were crossing

Ireland somewhere near Athlone, I believe. I was leading, going like a lunatic as usual and I rushed down this hill into a small hamlet and up the following hill, out and on for about another mile or so when I realised that no one was following. I stopped, hung about for a few minutes then went back! As I arrived at the top of the hill, I saw a large crowd with Symes in the middle, arguing with the locals. John appeared and explained that Richard had reshaped a gentlemen's push bike; he had apparently pushed it out in front of Symes whilst looking at my rapid exit through the village.

They were haggling over how much compensation Richard should pay when suddenly, the door of the local pub opened and a large red headed man strode over shouting, "don't give him nothing, don't give 'em nothing they will only waste it on beer," eventually things were settled for Half a Crown (2/6 or 12½ pence in today's money).

On another occasion, we came upon a large horse in the middle of the road that looked threatening, eventually Richard drove his Triumph at it, and it turned and ran off down the road. We visited the 'Camlin' at Crumlin, Near Belfast and Loch Neagh and saw a film about London's Tower Bridge which had a story wrapped around it, but I cannot remember the name of it. That is unusual for me as I can usually remember almost all films I have seen and the actors starring in them! I do remember that on several occasions the film broke, and a huge roar went up from the audience. When the interval arrived halfway through, we expected the usual ice cream girl to appear and when we inquired for refreshments, an order was taken, and a boy sent up the road to the local shop!

Continuing our journey into Northern Ireland, we were camping most of the time, but on this occasion we were offered accommodation on a farm by very friendly Scottish people. We were fed and entertained in their parlour very tired, but Richard had got the bit between his teeth and spent the evening talking;

asking the farmer, "how much are your rates?" whilst Anne, John and I had virtually dropped off!

Even further North, we came to the Glenshane Pass, over the Sperrin mountains. Now, at this point we come to a 'bone of contention'! Symes maintained that his bike had more power and was faster but for some time previously he had to run on 'Posh' shell petrol because of 'pinking' detonation, Whereas John and I used the cheaper Caltex with no problems. The Glenshane Pass resolved the issue once and for all, it was a long hard steep haul, but our Norton's took it in their stride and stormed to the top 1,500 feet in grand style. Eventually, Symes appeared slowly in low gear with his engine pinking heavily even on the Shell fuel. On into Donegal we went, the weather deteriorated. thick fog and/or low cloud arrived, and we got lost creeping over the moors for miles not seeing any life except several derelict farmsteads.

We finally arrived in the late afternoon at a small village which consisted of a High Street going uphill as you would expect! With houses/pubs on either side. Upon enquiry we were told, "don't bother to put your tent up, there are several empty houses up the top of the Hill, the owners have gone away to America!" So, this is what we did and the house we chose had more or less everything there just as if the occupants had got up and gone for a walk! Richard and Anne decided they had had enough for one day and retired, John and I decided we would do all the pubs in the village, one pint at each. I know we didn't achieve our aim but returned somewhat worse for wear and not very popular with the Symes'.

We explored Donegal's famous 'bloody foreland' so named on account of the deep red colour of the sands – Donegal at that time was very remote and underpopulated, no modern bungalows as now! Only the odd thatched cottage, the EU had not pumped vast sums of money into Ireland's economy; ruining the look of the place with tourism etc – the locals probably wouldn't agree, living without electricity, piped water, etc! And so on to Slieve League, these are spectacular cliffs on the south shore of Donegal which drop a sheer 1,950 feet into the ocean. In those days they were only to be reached by a cart track over which we crawled in first gear at about 2 mph. From here after a night's stop, we moved on into County Mayo and through to Achill Island, where to our surprise, we were greeted as we crossed the rickety old wooden bridge by people coming 'on mass' out of their houses and cheering and clapping as we went by! A lot of them, I am sure had never seen a sidecar outfit. the houses they lived in were little better than hovels, some of them beehive shaped!

We camped on the western beach, under the Lee of Croaghaun, a 2,000-foot, tooth shaped mountain for a couple of days/nights but the place was so remote that I don't remember being able to go for a drink! In those days, there were virtually no pubs in rural Ireland, all you had was a general store which sold everything (Arkright) with one end separated off, with the word 'BAR', which sold Guinness of course, and Phoenix beer, which was a bitter, no lager I am pleased to say! They came later with the troubles and the EU. We returned to Roslare hence Fishguard and set off in inclement weather for Wales on the A40. We got as far as Abergavenny or thereabouts when my gearbox gave out.

The large bearing supporting everything, including the clutch and final drive, disintegrated. I had noticed the last few days that the transmission had felt rough and investigation had shown that the clutch assembly was literally being supported, hanging between the primary and secondary chains! Eventually, the primary chain broke and that was that. Luckily, we had stopped near a garage so I asked the owner if he would mind if we made a few <u>minor repairs</u>. About seven hours later, after extensive grovelling, using spares from the Dublin, purchased spare gearbox, and some support from the garage owners' workshops, and frequent cups of tea; a lady over the road had noticed our predicament and bought over some lunch! We were able to continue home.

Vintage racing continued as before but 1962 was a disastrous year for me, as race after race I failed to finish, or finish way down the field, resulting eventually in engine disintegration at Snetterton. The problem was eventually traced to detonation (Pinking) leading on to pre-ignition, The result of too 'soft' a central sparking plug. one of the reasons that it took so long to find, is that I had no trouble at all during a successful 1961 season with the same setup!

It was at this time I bought my second Scott GRU 668 and was using it regularly for work at De Havilland's, unfortunately it was fitted with Dowty 'Oleomatic' Front Forks and they leaked, so that I was forever having to pump them up! My solution was to fit a pair of Norton 3 spring 'Webs' shown on the photograph, no problems after that!

Also, around this time I started to convert an Excelsior Manxman Engine, a 250cc four stroke to a uniflow scavenged, Rotary Valved inlet and poppet valved exhaust – two stroke! I machined it all up using my primitive workshop plus De Havilland's facilities.

It died, stillborn, I'm afraid; but it was to reappear later as a supercharged Triumph 500cc twin cylinder. The girlfriend of the time was Alison Soffe whom I met at Addiscombe nursery where Margaret Durrant worked before, she married Bob Durrant. We all went to a wild party there one Guy Fawkes Night; that is Ernie Evans, Jeff Gunter, John Lane, Clive Heasman, who later married Mary, and me. The problem with 'Soffe' as we called her, was 'marriage' and on one occasion she took me down to West Chiltington in Sussex to meet her parents. They lived in a very nice little thatched cottage and were nice people but seemed to view me as a prospective son-in-law! So once again, I ducked out of the situation and became single.

The 1962 annual holiday (We only had two weeks per year in those days) was again Ireland, only this time it was an all-male affair: Simon Grigson – 1938 Norton 500cc solo, Gerry Fitzpatrick – 1930 Brough superior 680cc solo, John Lane – Norgon Superior 1000cc- sidecar, and me – 1938 Norton 720cc – sidecar.

From the start, it was realised that this was going to be a 'thrash' as three of us were vintage racers! Simon and Gerry went on standard machines but with John and I, it was quite different. I modified the Norton engine by increasing its capacity from 500cc standard to 720cc, by fitting a Panther piston of 87-millimetre bore, with the long stroke 120mm Big Four Norton, side valve crankshaft, all of which proved entirely suitable for pulling a heavily loaded sidecar.

Meanwhile, in secret, John was building a 1,000cc monster, which utilised a Matchless Vee twin engine taken from a 1938 Brough superior and grafting it into an ES2 Norton frame etc, both of these machines can be seen in the Blarney Castle photograph. This same engine was fitted in Morgan three-wheelers in 1938/39, hence the name 'Norgon Superior'. Gerry and I set off first and had a fairly uneventful ride to Liverpool. John and Simon, on the other hand, had repeated seizures from John's bike because he had not had time to test it beforehand! However, they finally arrived, late, but we were told that the main boat doors had been closed and that all passengers etc could only board by gangplank. This was alright for the solo machines but a 'no go' for the sidecars. There was nothing for it but to separate bike and sidecar and manhandle them up the gangplank with the aid of some willing passengers and crew members, who thought it all a huge joke! Another dreadful wet/rough crossing to Belfast, it takes about ten hours.

The first thing I had to do was find a motorcycle dealer and purchase a new rear tyre as mine was on its last legs! this tyre was bald when I subsequently arrived home after 900 miles of thrashing! Our first stop upon leaving Belfast was Dundrod for the Ulster Grand Prix where we witnessed Alan Shepherd on a 350cc 7RAJS lead the world champion John Surtees on the mighty four cylinder works MV Augusta for the first few laps, until the gudgeon pin broke. Such was the performance that MV registered a protest, demanding that the AJS engine be stripped to ascertain its capacity. they thought it was a 500cc G50. An incredible performance!

From here, we went via the Antrim coast road to the Giants Causeway and then into the beautiful countryside of Ulster. It was in a small village, at one of our frequent stops for ale that we saw a poster on the pub wall saying 'seven nights of revelry – come to Scotstown etc' just the thing for us we thought! Scotstown is in County Monaghan. We stayed there for several days, camping in a field alongside a River. Very nice. We met several nice young ladies, some on holiday from England, plenty of drinking, dancing, merrymaking generally. We were well received, and our bikes caused a lot of interest.

Some further points of interest perhaps! Our cooking arrangements consisted of my faithful army petrol stove with a sheet of 16 SWG stainless steel over the flame to grill/fry steaks, sausages, etc. Very handy because we could tap fuel from any bike when necessary. I remember on one occasion passing a fried egg sandwich into John's greasy oil-soaked hands, as he was in the process of removing a cylinder head for an internal examination! And watching the yolk run out between his fingers! Yuck!

Sleeping arrangements were primitive, I had an American officer's folding bed an army sleeping bag but the others, I think just had 'Lilo' beds, except Simon Grigson who used the bread knife to dig a hole in the ground every night to rest his hip in! Our washing and toilet arrangements were also primitive, we used rivers, council toilets if available. I only took a toothbrush and hand towel with me for a fortnight! Gerry tried to keep up appearances by washing his trousers etc and hanging them on a fence; unfortunately, he forgot them (he had a spare pair) and his mother was not pleased when he returned. She said, 'they were a new pair, why couldn't you lose the old ones?" A look at the accompanying photograph of us will show quite clearly what I mean!

Returning to our travels, we were now heading west for Sligo and County Mayo etc, when we had two incidents. With John leading, at high speed as usual, we arrived at a hump backed bridge where he proceeded to take off, the resulting landing, collapsed the

101

assembly so that it folded inwards, trapping Johns left leg. Fortunately, without injury but it took a little time to extricate him and realign the outfit for further travel. The roads in Ireland at that time were generally atrocious, serious holes that could damage the bike if you hit one head on! You will hear what happened to my machine later on. The second problem occurred due to the constant hammering; I was in front of Gerry and Simon, both on solos if you remember, when the bottom of my sidecar body fell out! The heavy objects, like the spare gearbox, Magneto, tools etc all went bowling down the road behind me and when I looked over my shoulder, there was Gerry and Simon, swerving all over the road, trying to avoid these items, not knowing which way they would jump next!

Upon arrival in Sligo, we camped in a field at the back of some houses on the outskirts of the town. This proved fortunate in many ways because after a day or two we were taken 'under the wing' so to speak, of several housewives who insisted on cooking our meals and offering us proper washing/toilet facilities, great! Interestingly from our campsite we could see a large hill with something on top of it. Investigation revealed that it was a Knocknarea (1,078 feet) therefore a mountain and a blip on the top was the tomb of Maeve, Queen of Connacht. Interesting, we did not climb it, probably running short of time. It would be interesting to go back and take another look at it! This said in <u>2014</u>.

More drama occurred at this point, my sidecar wheel, which had been wobbly for some time, had several broken spokes and was no longer functionable! Eventually, a sweet shop was found in the town, the man being also a cycle enthusiast, cut some to length and I re-spoked the wheel, sitting on the steps of Sligo Town Hall with an interested crowd around me! From here, we continued west into County Mayo and onto the Mullet Peninsula. Belmullet, the capital, at that time was little more than a 'shack town' with wooden boards as pavements, almost like something out of the Wild West of America, that you see in cowboy films. The children who flocked around to see us, had nothing on their feet and had never seen motorcycle/sidecar machines! Turning south, we visited Westport House, where we made contact with 'civilization' once more, to find that 'Oh, Yon Railway Station' by the Temperance Seven was top of the charts!

Continuing South along the coast road, spectacular scenery between Louisburgh and Leemane, we arrived at Clifton in the Connemara County park, no national parks in those days! From here we visited the landing site of Alcock and Browns transatlantic flight in 1919, where they crashed in an Irish bog after being the first aircraft to fly the Atlantic Ocean. Next stop, Galway, camped near the beach, never used campsites in those days, in fact, I don't think they had even been invented.

The last time we had done some dancing, if you remember was at Scotstown, so we decided to visit the Galway ballroom. Simon was our star performer, he only had Wellington boots, but the girls didn't seem to mind! The rest of us were no better in riding boots etc, all of us pretty scruffy by now. The interesting thing here is that the sexes did not line up each side of the dance floor, as they did at the smaller events. Most peculiar as soon as the music started, they all used to rush across the floor at one another! From Galway into County Clare and a visit to the cliffs of Moher, which were just that! Nowadays, it's a tourist trap, with gift shops, toilets, charges to get in etc. Awful.

The next stop was at Ennistymon, where we camped in the old railway station for two to three nights. The lines had been taken up, but everything else was more or less intact so we just put our gear in the waiting room etc and made ourselves at home; the locals recommended it!

We explored the local area by day and spent the evenings, often until midnight, drinking and chatting (no licencing laws here) with the owner of the bar, a Mrs Carrigg and an old man who had fought the British Black and Tans in the 1920s when he was in the IRA. He told us that on certain occasions when being chased by the police, he had hidden in caves in the cliffs of Moher! He knew De Valera, the first independent Irish Prime Minister but didn't think all the troubles and killing on both sides was worth it – violence solves nothing as proved in 1968, etc. We went to the Dingle Peninsula with its High Peak of mountain Brandon and surrounding beautiful scenery, including excellent sea views of the Blasket Isles, which used to be inhabited, The old dwellings still visible from the mainland. It was here on the side of Mount Brandon, descending the Connor Pass we had our most serious incident.

I was in the lead at that time, the road was wet from a recent shower and Simon Grigson was right behind me trying to get past! Using all the road, with drops on the left and solid rock faces on the right, I was able to drift the sidecar outfit on the slippery surface, but no such luck on a solo. I heard a rending noise, looked over my shoulder to see 'Griggo' and the bike bouncing down the road behind me.

He was somewhat cut and bruised, nothing broken fortunately, and we spent some time in the afternoon patching up both him and the bike before finding a campsite for an early night stop at Dingle. However, after a good night's rest all was back in order. From here, we marched onto the Ring of Kerry and up through the gap of Dunloe, much to the annoyance of the gypsy horse travellers who seemed to think it belongs to them; to ply their tourist trade. A lot of the locals, disagree!

Another potential disaster was avoided by myself this time, I was pounding along trying to catch up the others, when at the end of a long straight, the road suddenly turned sharp left. There is no way I was going to get around the bend, luckily there was an open gate with a drive leading up to a posh house.

I shot up this drive, spraying gravel etc everywhere with the brakes hard on and just had a glimpse of a man and woman gardening, as I rushed by. I swung it round in front of the house and accelerated back down the drive, giving the two people a cheery wave – they looked astonished! Up until now the weather had been good but as we started to move east towards Cork, it started to rain, and we had a very wet night under canvas with water coming into the tent and running under the ground sheets. It was decided therefore, we would have a bed and breakfast in Cork and dry out all our gear, if possible.

We set off somehow separately and lost one another. I remember pounding down this straight road trying to catch up, when I saw a man come rushing out of a house with shaving lather on his face, waving his arms for me to stop! No way I realised he was a policeman who had been disturbed by the others rushing through at high speed and was determined to catch someone. I managed to swerve around him. Eventually, as I approached Cork, I caught the others up. Everybody seemed to think everybody else was in front of them, hence the catchup speed! Thick!

We managed to get a B&B in Cork at short notice, on the condition that we got up early the following morning, as night workers coming off shifts would require our beds! From here, we now went to Blarney Castle where Simon kissed the famous Blarney Stone, which involved laying on your back and hanging your head over a 50-foot drop, upside down – not for me! See photo of me and bikes. We had an uneventful ride back to Rosslare, hence Fishguard and driving down the A40 towards London again.

Since the incident approaching Sligo, when the bottom fell out of my sidecar and the resulting wheel rebuilding due to excessive speeding, combined with atrocious roads things had been getting progressively worse. The chassis had broken in two places, so we resorted to lashing it together with rope, some of which we had brought with us and some purchased on route. When we finally reached home the whole sidecar chassis and box, was scrapped! I remember burning the wooden boat in my dad's garden, I think the whole thing only cost five pounds, all of the sidecar boats were scrapped after these Irish holidays!

More or less as soon as we returned, we were invited to one of Laurie Ascot's well-known parties, he was another vintage racer with John Wilkinson, both were old Whitgiftian's, But slightly younger than me. I think they were in John Collett's class at Whitgift. We four went, plus Ernie Evans, Geoff Gunter and it was here that John met Lyn! Jerry had ideas of dancing with Lyn and asked if it was an 'Excuse me' to be told firmly by John that it wasn't! So that settled that!

After the last vintage race of 1962 at Mallory with yet another non finish, I was faced with a decision whether to do up the Norton, which meant considerable work or go for something that I've been thinking about for some time, which would involve even more work! See article in *Yowl*, Scott magazine circa 1963/64. I decided on the Scott/Norton and spent the very cold and long winter of 1962/63 in my father's garage building it.

Meanwhile I continued to go to De Havilland's having some horrendous cold journeys from Wallington to Stevenage across London. Coming home one night followed by a police car up Foresters drive, I turned off my lights as I entered Ingleby Way, and roared down my parent's drive and around the house at the back, immediately stopping the engine to peek around the corner to hear the police say, "he must be round here somewhere," another close one!

The real winter started boxing night 1962 and continued without let-up until the middle of March 1963! This winter and that of 1947 were considered to be two of the coldest and longest of the 20th century, in fact, in 1947 coal stocks ran out and a national emergency was due to be declared! Everywhere was so icy with continuous snow that I fitted a sidecar chassis to my Norton, in order to stay upright.

One particular journey springs to mind; I used to set out at about 7:00 am to arrive at work at 9:00 am – 2 hours, on this particular occasion, I set out at 7:00 am as usual but didn't arrive at

Stevenage until 1:00 pm in the afternoon! Things were bad enough getting through London with the traffic and all sliding about everywhere but when I got to the A1 North of Hendon, there was nothing! Just a whiteout! I could barely see the road; no other vehicles were moving and those that were there were abandoned! I was frozen, I made one stop at Hatfield in a cafe for a cup of coffee, then carried on. I think I only saw one other car going the other way in all that time. The worst ride I've ever had, I could barely get off the bike when I arrived, the lads in the office said I was mad to have attempted it. Still, I don't suppose it would have been much better the following day/week etc.

Things got so bad the car parks at work were deserted; only my bike and one or two others got through, almost everybody walked to work and, in the evening, when it was dark you would see a torchlight procession all the way up 'Six Hills Way' to Stevenage new town about a mile long, incredible, never seen anything like it since! It went on for weeks, all through January and February. I used to pass rows of lorries at the roadside with fires burning underneath them as the drivers used to try to thaw out the frozen diesel fuel in the tanks! Don't try it with petrol!

As things eased up, a trip was arranged for the Stevenage mob to meet the Croydon lot at a halfway point in a Hendon pub where Evans, Gunter, John Lane, Hamilton, Symes, etc, met Jerry, Snook, Rex Boyer and myself – we went down in Jerry's open top car, a cold ride. Around this time, we were invited to a party at Ron Creswell's house at Hennicker Road, East London. I went with Bob Hamilton, Roger Stanton, Ernie, Jeff Gunter, Gerry Fitzpatrick, Ascot, Wilkinson and several other vintage racers including Guy Sindon, Bob Burnett, ex Whitgiftian and others. It was a great do, we all got sloshed and went round to Ron's local pub just down the road where they had a Penny Farthing bicycle leaning against the wall.

I said, I'd like to ride it, so out into the road we all went, and I managed to get on it and rode it up and down the road without falling off, first and last time I ever rode a Penny Farthing – great fun. Back to the house where we continued drinking – I gave a speech on the American Civil War if I remember correctly. Ron Stanton had a narrow escape when he stepped out of Creswell's barn upper floor, thinking that's where the stairs were, somehow, he landed on his feet in the cobbled yard, shows how far gone we all were! I don't remember how we got home or who drove – this was long before the dreaded breathalyser!

Vintage racing had recommenced by now, with the first meeting of the year at a cold Snetterton – It was always cold at Snetterton! So, I took the Scott/Norton for its first outing. It was a disaster, a holed piston halfway through the race at the end of the Norwich straight, due to a weak mixture of methanol.

Brands Hatch was next, brilliant! A win, no problems. Back to Snetterton and again a ruined piston, same place. The carburettor set up I had was wrong (see notes); and with a revised set up the problem disappeared, resulting in more leader board positions, and finally winning the coveted Rex Judd National Vintage Road Racing Trophy for 1963! The prize giving event was held, I think near Rugby and was attended by all those shown on the photograph, plus many others i.e. Roger Cramp, Clive Waye, John Wilkinson, Frank Booth, Ivan Rhodes, Dave Lecog, John Hurlestone, Simon Grigson, Sandy Grigson, Dan McDiarmid, racing secretary and of course many others. The girlfriend at the time was Carol Roach, sister of Barry who also cycled to Whitgift with Symes and me each day. It was noted by Sandy Grigson, whilst I was 'jiving' that 'I was a lovely mover' that's called 'taking the Mickey', we still laugh about it!

It may be worth noting at this point, that there was some controversy over the whole Scott/Norton episode! To recap: when I came up with the idea, I put it around generally amongst the vintage people, especially the VMCC management that I was thinking of building a 'special' but of course, it would consist of pre-1930 parts. Nobody really objected. I don't think they thought

I'd do it, and all would have been okay if it had not been so successful. But when it started to win the grumbles began. Not, I hasten to add from the riders, but from the vintage purists. However, there was a Cotton/Norton ridden by Jack Squirrel, even Clive Wayes winning Scott was a mishmash of bits; no Scott frame was ever quite like that! Anyway *fait accompli,* that was it. I had Titch Allen, founder of the vintage motorcycles club and ex-racer on my side. he was a lover of specials and said on one occasion that he thought it was, 'the most successful vintage special ever built!' I understand, nowadays but they will not allow any new specials to be built and raced but existing ones including my Scott/Norton are acceptable, it's over 50 years old now anyway!

It was around this time I became unwell and had several bad migraines at Church Lawford etc; I even raced whilst having one with only partial vision! My condition got worse and I went home on sick leave from De Havilland's. I was in fact having a nervous breakdown, but our GP was out of his depth with mental illness as are many people today but there is more awareness of it now. I was hanging around at home, suffering panic attacks, some delusions, lack of concentration, sleeping badly, loss of confidence and also physical effects. I would come out in a great sweat at even small exertions like pushing my bike to start it. It was at this time that Jerry and Ted Snook suggested a holiday on the continent, and I thought it might snap me out of it, so I went along. We went in Ted's Dormobile van with the usual camping gear and went South through France into Germany. I can't say where exactly we went, I remember we arrived at Koblenz on the Moselle which is a tributary of the Rhine and drank 'Schnapps'. We returned, as far as I can remember, back through France, across the Massif Central, climbing to over 4,000 feet in mountainous regions, all very exhilarating I am sure, but I was too ill to appreciate it.

Upon arriving home, I was taken to Fairdene Mental Hospital, where I resided for about two to three weeks and met some interesting people, both staff and patients. A Doctor Raymond, a man of about 60 years was my psychiatrist and he realised that basically there was nothing seriously wrong with me that time and rest would not put right, so apart from putting me on 'Largactil' and sleeping tablets, he gave me no treatments at all, they did give me a brain scan, but it did not show where my lunacy came from! He did, however, offer me some good advice. If when applying for a job or whatever, do not tell them you've had a breakdown, unless they ask directly and do not be inclined to make monetary donations to organisations specialising in mental health!

I was an ebullient patient and some of the other patients used to ask me why I was in there as I didn't exhibit any particular symptoms! I wonder what they would think now if they read these memoirs! Eventually I got to thinking that while I am mooching about in here, my preparation for the defence of my vintage road racing trophy was slipping by, So I had my parents bring me a Norton cylinder head which I proceeded to shape and polish. But it was no good, I needed more tools than were available, so back home I went. It was a mistake, I went back to work at De Havilland's but after less than a week, I was back home again having suffered more severe migraines and delusions etc. I was unable to concentrate on my work and was hesitant about making two and two add up to four! Shows what a state I was in! It was awful, not to be recommended to anybody.

The result was that Doctor Raymond put me on a course of outpatient visits at the hospital And I used to attend group therapy twice a week where we all used to sit around and discuss our problems. It worked for me because I began to understand something about mental illness, which I was totally ignorant of before and to also realise there were many different types, for which in many cases there was a cure or at least alleviation. Essentially, I was told that mine was a neurosis rather than a psychosis. the latter type being the more serious. Sorry to have laboured this topic but it is the first time I have ever written my feelings down; always kept them more or less to myself, until I married Jill (nurse) and came into more contact with some of the problems of her family.

It took many years to finally shake off the effects of this breakdown, as occasionally, for no reason whatsoever it seemed, I would experience a strained feeling behind my eyes (difficult to describe) and a slight panic attack, just enough to tell me that the illness had not finished with me just yet! Upon reflection, I think what helped me most was having a strong interest in engineering; in fact, it's been my whole life really, in one way or another with my jobs and hobbies etc, such that I've never had time to be ill, if you see what I mean! And I am sure that this attitude helped to bring me out of the depression etc. A purpose in life. Hope it works for you!

Stevenage 1961 at my digs with Mrs Ross and her six sons.
My second Scott whilst working at De Havilland Propellers.

Hastings Aircraft CIRCA 1962
completes a long job

After almost daily flights from Hatfield for eleven years the familiar four-engined Hastings is to be returned to the Ministry of Aviation. Since 1951 valuable research into de-icing problems with various propellers has been carried out with the aircraft by the Manor Road Flight Test Department headed by Mr. Cliff Aubrey. In this photograph of the Hastings, landing after a recent sortie, the spraying rig on the front of the fuselage can be seen. The familiar roar of the Bristol Hercules engines during run-up will be missed by many people at Manor Road.

On John Lane's 1928 racing Norton circa 1958 taken at 76,
Queenswood Ave, Beddington, Surrey.

VICTORY FOR VINTAGENT

R. G. " Bob " Collett, of the D.T.A. Stevenage, crackled round the
Brands Hatch circuit on Sunday, June 25, to win the Vintage race on his
1929 Norton. The machine, a model 18, 490 c.c. Norton is tuned and
maintained by Bob and, as his 60·38 m.p.h. lap average proves, he does
it well.

Year - 1961. 1st Win.
De Havilland Monthly Magazine

Rob Collett, Richard Symes, Ernie Evans and
Anne Symes on holiday in Ireland - 1960.

Scott / Norton - 1962.

Gerry Fitzpatrick, Simon Grigson, Rob Collett in Ireland - 1962.

'Norgon' Superior, V twin 1000cc Matchless engine in Norton motorcycle
with my 720cc Norton at Blarney Castle, Ireland - 1962.

Chapter Seven

Agi – 1963 to 1966

As a result of this set back, I resigned from De Havilland's and spent some time at home, eventually in early 1964 getting a job however at Aeronautical and General Instruments (AGI) in West Croydon. I was employed as a design draughtsman working on the cameras of the TSR2 Britain's leading Strike Fighter/Bomber called a flying electronic circuit board.

The office was set away from the main factory by about two to three miles in a large old Victorian house in West Croydon. There were some interesting characters here; Miles Handley for instance, who was one of the first to design and fly his own hang glider and was a leading light in the formation of the microlight society. Hut was slightly more infamous; his claim to fame was the setting up each morning of a telescope trained on the house opposite, where a young lady used to get dressed! There was usually a queue at about 9:00 o'clock, and another character who claimed to have invented an electronic clock (not new now of course) and was going to leave the firm because he was worried that they could pinch his idea and patent it for themselves, as it was in their line of business! This was not entirely unreasonable, as in those days there was very little protection for inventors; it was seen by organisations that, when employed, they owned you 24-hours a day or you had invented it in their time! Things improved in the 1980s but more on this subject when we get to my Hunting's period.

Clive Woodfine, John Lane and I used to meet at the Windsor grill, a cafe outside West Croydon station every working day at 12:30 for lunch. John worked opposite and because the AGI works canteen was at the main factory we were given luncheon vouchers, this went on for about two years and Clive is still a lifelong friend, we exchange Christmas cards!

It was during my time here that the idea of the Excelsior Maxman was revived, as a blown 500cc twin cylinder, 2 stroke, based on a Triumph Tiger 100 engine, using a 750cc Sharrock's Supercharger! Large support was given by AGI, in the form of Con Tee, the workshop foreman, who later came and gave support when we took it to Croydon Aerodrome Perimeter track and in a modified (lengthened) Norton Frame (see photographs) got it nearly to 100 mph and 5500 rpm with brisk acceleration! Astonishingly, it started third kick and required very little jet changes to achieve stable running, an article in the *Motorcycle* was written by Vic Willoughby their technical editor in June 1965.

Life at home continued with Phyliss introducing television, initially against our parent's wishes. They read or listened to the 'wireless' but once it arrived and was installed in the back room, they were gradually 'sucked in'. My mother meanwhile had been grousing about the 'rubbish' as she called it which was in a heap outside the French windows in the back garden. She said it was an eyesore and when she had visitors in for cakes it was an embarrassment! I tried to explain in simple terms that these bits i.e. frames, forks, wheels and subsidiary cycle parts were valuable vintage items and were being kept as spares etc, should anything go wrong with the racer or on the road machines. In all fairness, she repeatedly warned me and one day upon returning from work they were gone! She had lost patience and phoned the council who had removed it; you could get things like that done free on the rates (council tax) in those days. still it was my own fault!

I spent most of my evenings after about 8:00 o'clock with the lads down at The Plough in Beddington just at the bottom of John's road. It is still there and is a famous old landmark. We drank and played snooker in the backroom, with me occasionally taking time out to leer at Iris through the window; where she was most evenings in the saloon bar with her 'fellowship' friends and boyfriend, John Clayton, with whom it was said, she had been going out with for ten years! Apparently, the problem was that he wanted to marry her and move her into his house, with his mother! Imagine, two women in the same kitchen! My progress with her was slow, until one evening Bob Durrant informed me that she was having problems with her car starter motor and he had recommended me as a suitable mechanic! From this point, things speeded up and after an altercation with Clayton outside the pub, where he threatened me with violence and later outside Iris's house, when for some reason he insisted on seeing her father, I don't know what he expected but the old chap wasn't interested, understandably and threw us out telling us to sort it out! Iris, I think went upstairs to bed and that was that end of the 10-year engagement!

Our romance had now started apart from one hiccup when she was on the train with Phyllis going to work in the city, when somebody informed her that, 'a car like yours is going along the road parallel with us',

'What? That is my car!'

I was using it to go to work, having told her that it was off the road while I continued to work on the starter motor, I can't remember what excuse I used to get off the hook, but I know all the lads had a good laugh!

At this time, we could 'burn the candle at both ends' and come and go at all times of the day or night and I remember, on one occasion, when I had a particularly hard night creeping in, in the early hours and getting up at 5:00pm in the afternoon just as my father was coming home for his tea; I was just coming down for breakfast! My mother, quite rightly, said I was using the place as a hotel! This was a typical incident we sometimes got caught up in; remember parts of south London were quite rough. We were sitting in the 'Criterion' restaurant in South Croydon when a fight broke out at the next table and things started to 'fly about'! Our table was jostled but we just lifted our plates and carried on eating, the staff appeared as if from nowhere, police arrived, some of the combatants were now fighting in the street outside. when order was finally restored, Hamilton admitted that he had kicked somebody's shoes across the room and to another table and one of them had accused the others of trying to pinch them, most of them were the worse for liquor! Stupid boy!

Moving on to more intelligent things, AGI were an optical firm and made cameras etc, grinding their own lenses, eyepieces. Etc. And I managed to scrounge a four-inch diameter parabolic mirror and eyepiece, which I fitted into a cardboard tube, making a very respectable telescope with a focal length of about three feet (one metre). More powerful than anything I had looked through before. By this time, the Isle of Man TT races had arrived, and John Lane went more or less religiously every year. Richard Symes, Jerry Fitzpatrick and I had never been, so we decided to go this year – 1964, hoping to see the two stroke MZ's dominate the lightweight classes i.e. 125 and 250cc. We all went on bikes; John – Norton, Richard – Triumph Speed Twin, Jerry – Brough Superior and me on my tatty old 16H side valve, Norton.

Trouble started on the docks at Liverpool when a policeman noticed that I wasn't displaying a tax disc. Well, I explained that I hadn't got one but was going to apply as soon as I got home! This did not satisfy him but as he pointed out he could not prosecute me as the docks were not the public roads but private property, Isle of Man Steam packet Co Ltd! And furthermore, if I took it to the island, I would become subject to British road regulations again. Impasse!

Now I suggested there were two options: -

1. I take it to the Isle of Man, but do not ride it on the roads, just leave it at the dockside and go on one of the other lads' pillion.

2. Throw it, here and now into the sea. This I was prepared to
   do as the bike was a bit of a wreck anyhow and go home on
   a pillion after the TT. So, he backed down, things were
   getting too complicated and with the boat about to finish
   loading, I started to walk away telling him to push the bike
   over the side!

I therefore ended up taking the bike, riding it round as normal, he
had said he would phone up to make sure I left the bike in the
docks but nobody appeared so that was that! We had a good time
there; I shared a bed with John and Richard and Jerry slept
together; unfortunately, we were some years too early and the
four-stroke domination continued; Honda's and MVs doing the
winning.

A change to my normal racing programme occurred here; John
had entered for a grass truck meeting at Casiobury Park, Watford
and could not ride. I think he was recovering from the effects of
glandular fever. So, in order not to waste the entry, I said I'd have
a go! Well, riding his grass track, knobbly tired CSI Norton; I rode
like a demon and finished second in the first heat and second again
in the final! I spent considerable time on the ropes and almost
collided with the other riders, that at the end of the proceedings,
Willie Wilshire, president of the grass track section of the VMCC,
was also riding and asked me if I was thinking of taking it up
permanently; my negative answer gave him much relief!

Also, this year I started to race Gerry Fitzpatrick's 680cc Brough
Superior (see also – racing Brough Superior article – to go in Titch
Alan's book on Broughs). First meeting at Cadwell Park in May, I
was second in spite of a slow speed crash at the hairpin which
wiped off the oil pump on the last lap, so I picked it up, grabbed
the oil pump and pipes, jammed them between my knees and so
finished the race!

I had no problems next time out at brands and finished fourth; on
the strength of this Gerry decided to uprate the engine to 1000cc;

a mistake! The bike now became too heavy and difficult to handle on the short circuits. I was knocked off it on the first lap at paddock bend at Brands by Mick Broom, he slid off taking me with him, the race was subsequently won by J. Lane (see photo) his only road race win, I believe and richly deserved too! Later, when at home that evening and watching the television sport news it said, "here are some of the highlights from the motorcycle racing at Brands Hatch this afternoon," and we all watched my crash all over again! Dad was not amused. After a further outing at Church Lawford without success, Gerry refused to put it back to 680cc. I quit and the riding was taken over by Ted Snook, he raced it through 1965/66, eventually giving up in despair as Gerry refused to alter anything on the bike, saying, "it was alright for Rob Collett, so it should be alright for you!"

Rex Boyer, at this time also decided to join us and bought all of Symes's model 18 Norton's; he had little success, with repeated mechanical failures, he would insist on under gearing and over revving! I used to tell him that the Norton engine preferred to pull rather than rev, but he insisted on exploring the five-figure rev band without success. Snook also rode his bikes and did much better by following my advice! Which was obvious anyway. I rounded out '64 by winning the Rex Judd Trophy for the second time, mainly due to the Scott/Norton.

My only two stroke car arrived about this time, it was a 998cc DKW (Auto Union) For which, I paid £100 from a man in Lincolnshire. It was of three cylinders with a freewheel mechanism to stop irritating 'snatching' during overrun due to fuel entering the cylinders from the idling jets, I never used it, preferring to change down and de-clutch, like one does on a motorcycle! It was a good car, very posh with a built-in radio, run on a 16:1 fuel/oil mix. It was a bit thirsty as you would expect with impressive brakes, as engine braking was poor.

500cc Triumph 2-stroke. First version as tested at Croydon airport - 1964. In lengthened Norton frame with single shorroks and 750cc supercharger giving +5 P.S.I. boost pressure.

Second version as a Sprint bike shown also in R. Bacon's book on the Triumph motorcycle. Boost pressure now + 15 P.S I. Note: third version made but no photos available - see text for details.

500cc Triumph 2-stroke as used in first and second versions. Jack's drawing.

1964

Hey! This is no life for an old chap!

Well! I'm 36 years old, y'know. We Brough Superiors were built to last, of course, but this is ridiculous. That youngster, Ariel, behind me,—now—he's only 34 But he's only a 500. I am (puff) 1,000. (980, to be precise.)
And here we are, at **BRANDS HATCH!**
This, they say, is Druids Bend.
As you can see, my breath is misting up a bit, but that's only to expected.
This chap who's driving—R. G. Collett, he's called, seems to be doing all right for a novice. After all, that's what the race is for.
And the fellow behind on Ariel. He's J. A.

Iszard.
I suppose this can be quite good fun (puff) —if you don't do it too often.

No, we didn't do too badly, at all. We were **FOURTH.**
Ariel nipped past before we could get to the flag,

I'm afraid. Caught us napping.
Still, a Brough Superior is something rather special.

125

Lovely sight for the vintagents—John Lane on the way to winning the vintage race on his 1928 490cc Norton. *1964*

Bob Hamilton at front. L to R - Bob Durrant, Margaret Durrant, John Lane's partner, John Lane, Phyllis, M. Lane, R. Collett, Carol Roach , Richard Symes, Ann Symes and Clare. V.M.C.C. Annual dinner and dance - Rex Judd trophy winner - 1964.

My wedding to Iris - 1966 at Saint Marys Church, Beddington.

Chapter Eight

1966 to 1968 – Sandall Precision Ltd

1965 was a preparation year for marriage, as Iris and I had now got engaged, at a party held at Iris's parent's house in Beddington, where both sets of parents met for the first time after which we set out on our holiday in Yorkshire (See photos) touring the Dales on Richard Symes' Triumph Speed Twin Combination, camping as usual.

We started looking around for a suitable house (starter homes, they call them now) but it soon became obvious that prices in the Surrey area were about 1/3 higher than the rest of the country! We could get much better value for our money if we moved away.

It was therefore decided that as I had considerable knowledge of the Hertfordshire/Bedfordshire area due to my previous working for De Havilland's and Iris had no objections, we transferred our attention to this area.

We finally settled on 18 The Pyghtle, Turvey, Beds, At £3,750 with a 6¾% mortgage, later to rise to 16½% in the 1970s, my father helped by lending me £1,000, Iris and I managed the deposit, she selling some jewellery and me selling my 7R AJS 350cc racing motorcycle, which I had bought about a year earlier for £250 intending to go 'modern' racing, but after a crash at Druids on the Brands Hatch practise day, I thought, 'if I damage this thing badly, it's going to cost serious money to repair'. So, I went back to vintage racing, of which I did not do much in 1965/66 for obvious reasons. I was due to leave AGI at the end of the year and I put my name on the executive register at the Labour Exchange (now Jobcentre Plus) to find work in the Bedfordshire area and as a result of this I received a phone call one afternoon asking me to meet a Mr A. Barker that evening after work at Croydon Airport lounge in the old control tower. The interview went well, and he said he would like me to visit the works when I was in residence at Turvey, for a further interview!

So, I left AGI, had a big leaving do at a pub in Duppas Hill, Croydon with the lads both from AGI and my own mates, sorry to leave! About this time, end of 1965, John Lane married Lyn down at her home in Bournemouth. We all went, Iris, Ernie, Gunter, Symes, Heasman, Durrant, me, etc, and had a great time. John got hopelessly drunk at the stag night due to Lyn's brothers and father mixing his drinks, which we thought was pretty awful. From here they moved to Lyn's flat in Caterham, where they met Jill and Ray Davis who were newlyweds themselves! Meanwhile, Iris and I had got married in March 1966 (see video of wedding) at Saint Mary's Church, Beddington, With Jerry Fitzpatrick as best man, plus five bridesmaids – including Phyllis, Lyn Lane, and about 40 guests. The reception was held in the restaurant above the Gaumont Cinema in Sutton.

We then moved to our house in Turvey, where myself and Williams (the man next door) proceeded to lay turf, plant trees, build walls, concrete a sideway for carport etc. As the properties were newly built, the builders had left most of their rubbish behind as usual.

Iris was first to get a job, she put on her 'glad rags', fancy hat, etc, went into Bedford, came home to announce she was now secretary to the manager of the Cosmic Crayon Company! I Meanwhile, was awaiting an interview at Sandall Precision in Bletchley, Bucks for the position of Development Engineer once again, (see CV). It arrived and I was interviewed by a Mr Saunders who showed me around the works etc (nice man) And I started work almost immediately.

My parents along with Phyllis, visited on several occasions, staying at the Laws Hotel in Main Street, Turvey, whereas the

Hancock's – without a car, had to be fetched from Beddington and stayed with us, a round trip of about 200 miles across London, no M25 in those days!

A lot of it was done in Iris's Austin 10, that I had repaired in 1964 remember! We also visited Lyn and John at Swingate Farm, Walderslade, where they had moved after the flat in Caterham and had a 'hairy' moment when descending Bluebell Hill into Walderslade, even in bottom gear with both foot and hand brakes full on, the car was running away with us and I was looking for an 'exit' road praying that nobody in front of us would stop! (Cable breaks, like the Austin Seven, useless). after this and several tricky trips across London, I sold the car and bought a Hillman van to continue vintage racing, Iris was not pleased and still blames me even to this day!

1966 was a thin season with one victory at Llandow, Wales in appalling conditions; the rain was so heavy I could hardly see where I was going on a track that I had never been to before. We went on our honeymoon to Ireland, again on Richards Triumph and this time I concentrated on exploring the centre (Bog of Allen) Where it was less popular. One incident springs to mind: we had camped on a hill in the middle of a field and upon cooking our breakfast the following morning, we observed a large horse galloping along the road below, on seeing us, it changed course leapt over the gate and started up the hill towards us. It was, to say the least, a tricky situation. I went to the bike, turned on the ignition, waited until the animal was about 20 feet away and depressed the kick start. it was a good starter, luckily, and when it roared into life the horse reared up, turned through 180 degrees, and disappeared down the hill, re-jumping the gate and off down the road! Goodbye! Ask Iris!

On another occasion, we camped again in a farmer's field and as it grew dark, lights came on about halfway up in the surrounding hills and we thought that there was probably a hamlet or village up there but, in the morning, even with binoculars we could see no evidence of habitation! Strange! The lights did not move, so it rules out people. The centre of Ireland was very rural in those days, many of the villages had unmade roads with boards for pavements and every other shop seemed to be a bar!

The job at Sandals was very interesting and I worked directly responsible to the managing director 'Alfie' Barker, he was also the owner. He was a man in his 60s and a bit of a tyrant who interfered in everything And I mean, everything! Even the colour of the toilets! He was a self-made man; credit to him, he was also in the right place at the right time. What happened was, he ran a small jobbing shop just before the war and when war was declared, all these small manufacturers were co-opted to start making items for the war effort and government subsidies started rolling in – full employment for all! So, his business and premises started to grow. After the war, they started building new towns like Welwyn Garden City, Stevenage, Newtown and Bletchley, etc, and in order to encourage people to move out of the overcrowded cities like London, Birmingham, Coventry, etc, jobs had to be created. So, firms like Alfie's, were encouraged to move out, heavily subsidised, both for the people and employers. As I say, he was a tyrant and interfered in everybody's job, including mine. The trouble was, although he was a good machinist man and could work any machine in the place, he was no designer or professional engineer, having no technical qualifications i.e. a self-made entrepreneur!

I later discovered that the man I had replaced, Mr Selwood had left hurriedly, after a blazing row with 'yours truly'. The poor man had much more to lose than me as he was a University graduate who had been enticed away from a very good job, by extensive promises from Alfie. he later had a nervous breakdown from all the stress of it. Most of the staff, about 200 of them, has come up from London with him or were 'yes' men, who all wanted a quiet life. When we had a technical meeting once a month, held in Alfie's office, I found myself, like Selwood, speaking out and usually contradicting him, which inevitably meant that the meeting ran over 5:00pm and I came in for a lot of 'barracking' from the staff, who wanted to get home on time! Typical.

I have talked a lot about Mr A. Barker because I quite liked him, and he was a great character. I've seen him walk down the machine shop floor among all the 'grot', dressed up to the nines! In all his finery, suddenly stop, take off his coat, roll up his sleeves he would be operating a lathe, mill or something, explaining to the operator where he was going wrong! Whilst the other members of staff would be remonstrating with him to go to some meeting or lunch etc! My main job here was to develop a machine called the 'Ramsey Interface Detector' one example of which was installed in the oil pipeline at Esso's West London terminal, another at Shell's terminal on the Wirral in North Wales and a further one at BP's terminal in Hamble Sussex. The idea invented by a Mr Ramsey was a hydraulic Wheatstone Bridge, in other words it operated like an electrical Wheatstone Bridge; whereby, the viscosity of the fluid flowing through it was balanced – equal – but became unbalanced when a different viscosity fluid arrived and flowed through it sufficiently, until the status quo was re-established. Clever!

These things were in operation before I arrived, but they were still in the development stage so, as the various fluids i.e. naptha, paraffin, diesel oil, petrol, gas oil, etc, are all pumped along the same pipeline, one after another, so there will be an interface where they intermingle with one another, this is what the machine picks up as described and displays it on a rolling trace record.

These fluids are all pumped from say, Southampton to London, a distance of about 85 miles and in order to get them to move over that distance in a three-foot diameter pipe, with a pressure of 1,000 PSI is required. In order to return our sample back into the pipe, Alfie had developed a three-cylinder high pressure pump with special seals and valves to cope with the thinnest fluids like naptha and petrol, which are also highly abrasive. The valves particularly, were very difficult to get at and repair so I designed a revised system whilst Alfie was on holiday in Jersey, his summer home in the Channel Islands. When he returned, instead of being grateful he went into a rage because I had dared to change his design, this I think, was the final straw leading to my sacking! Just like Selwood.

Alfie, to his credit supported various entrepreneurs; one chap whose name I forget, had invented a sweet wrapping machine which used to make a great clutter when he switched it on. Another was a 'mile to the gallon meter' that Ramsey had invented, this was before they were done electronically and now fitted to some cars etc. The main line of work that bought the money in was a contract with Westland helicopters at Yeovil, Somerset to provide hydraulic systems for their aircraft. It was at this time 1967, that I was to become a member of the 'Institute of Patentees and Inventors' no.7334. (Ref certificate of membership).

This came about after I submitted a design for removal of the connecting rod by putting the crankshaft directly inside the piston, I called it a 'slider crank'. It is applicable to both two and four stroke engines, so much so that I obtained two Volkswagen air cooled cylinder heads from Clive Waye, With the idea of making a flat 4 engine with only two 180 degree opposed double acting cylinder's! Nothing however became of this as at this time I did not have any real workshop facilities. However, as mentioned, the idea is also suited to two strokes and some drawings were done for single cylinder operation, using the underside of the double acting piston as the pumping chamber, like the crankcase in a normal two stroke and a supercharged double acting one stroke, where both ends of the piston are working cylinders, firing every 180 degrees of crankshaft rotation; with normal port operation or uniflow scavenge with poppet or sleeve valves! An interesting series of projects for some enterprising chap! This was the start of a series of patents I obtained later when working for Hunting Engineering in the 1980s.

My mode of transport at this time was the Hillman van, whereas Iris went to work in Bedford on the bus. Apart from the Scott/

Norton Racer I was without a road motorcycle; all the rest of my old Norton rubbish had been sold or disposed of by my mother! So, I looked in the local paper, and discovered that a model 19, 600cc Norton of 1957 vintage was for sale in Elstow. When I arrived, a man was arguing with the owner because the bike was a non-runner due to a lack of compression, the man had dropped the price to £10 but it was declined, I offered him £7.50 'sold as seen', He accepted. It turned out to be a stuck exhaust valve, which was easily remedied, and I used it then for 17 years! Before selling it to John Martin at Hunting's. It was decided to return to Ireland again, Iris's second visit, this time borrowing half of Richards speed twin i.e. the Jet 80/sidecar. (See photographs and there are also some Lantern slides.)

We sailed from Holyhead to Dublin and set off westwards through the Curragh, the famous Irish horse racing centre, heading for the Northwest of Donegal. By the time, we arrived and after some camping stops the weather had deteriorated so we had a bed and breakfast at Gweedore on the 'Bloody Foreland', So named because of the red colour of the Atlantic coast sands! The place was really ancient and reminded me of *The Old Dark House*, a film of J.B. Priestley's novel *Benighted* of which I have a copy, although at the time I had no knowledge of either! Two or three very old people were running it and unlike the book/film, were very nice. I went through there years later with Jill and Jack, but it had all disappeared, I think the house had been pulled down; what a shame!

We worked our way South from here through Donegal and as you can see from the photos the weather now took a turn for the better. We camped just outside Killibegs on a farm track this time and Iris loved it because twice a day at milking time, a whole herd of cattle came along (see photo). From here to Sligo, where we camped on the beach and had it all to ourselves, and this was over the spring bank holiday weekend! (Ref photos.) On into County Mayo where we camped by a beautiful river, again with cows coming by, then we went through Connemara, the Twelve Bens, views across Clew Bay and the islands of Irish boffin, Irish free and Inishbeg; Iris later painted the Twelve Bens from a photograph and it now hangs in our garden room in Stoney Stanton.

Our final bit of spectacular scenery was on the road from Louisburgh to Leenane, past Mwealrea a 2,500-foot extinct volcano! To the Asleagh Falls. We returned home via Roselare to Fishguard and had a dreadful journey, heavy rain all the way; we were completely soaked to the skin upon arrival at Turvey.

Meanwhile, 'back at the ranch', Gerry Fitzpatrick, my best man, remember, was in the process of re-installing the Triumph two-stroke engine with his own designed frame, with the idea of sprinting it. He added a second Shorroks Supercharger at the rear of the engine and a second article was done by Vic Willoughby in *The Motorcycle;* unfortunately, he did not mention Jerry enough; I think his mother got upset blaming me but Jerry was there and only had to speak up; this and one or two other incidents was the cause of our breakup; he is the only friend that I have ever completely fallen out with, such a shame. I'm sure his mother was the main cause! Only child. However, when finished we took it to Tempsford Aerodrome which was deserted but still enough of the runway left for a decent run.

The boost pressure was now 15 PSI (one bar) And on the first run As I opened it up; at about 50 mph I was suddenly enveloped in flames, I snapped the throttles shut, banged on the brakes and the flames went out. Puzzled, I opened the throttles and was immediately enveloped again! The cause? We had forgotten to adjust the blowoff valves situated in front and behind the engine for the increased boost pressure and the rear one was ejecting mixture over the rear hot exhaust pipes, say no more! However, it was decided that the times were not competitive, and more work would have to be done; but Jerry, now uncooperative lost his enthusiasm and later returned the engine etc.

In 1981, Roy Bacon's book on the Triumph came out and the bike is described in a specials section. Vintage racing continued through 1967 with the Scott/Norton gaining two firsts at Brands Hatch and Cadwell Park and various other places. Back at Sandall Precision, I was still battling with Alfie and yet another nail was about to be driven into my coffin! I will explain. Jerry, at this time, was unemployed and to try and patch things up a bit, I suggested he might try for an electrical engineers' job with us, as I knew that they were looking for one. So, I went to see Alfie and he agreed to an interview, I told Jerry not to worry as Alfie knew very little about electrics and with his experience at De Havilland's etc he shouldn't have much trouble. A date and time were arranged; next thing I knew was a summoned to Alfie's office – where is Mr Fitzpatrick!? How embarrassing, what was Jerry's excuse? He didn't think he would like the job after all!! But he didn't have the decency to tell me! So that was that.

By now the end of the year was approaching and on 8 October, Miles was born! At Bedford hospital, North wing and a few weeks later Alfie fired me. It happened like this; I was called into his office at about 10:00 am one morning and informed that he would pay me one month's salary and wanted me off the premises immediately. Well, I said, that is impossible, I'm here on my bike, there is no way I can carry all my books, drawings, utensils, etc, so he said he would lay on the works van to follow me home and that's what happened.

Iris of course had given up her job and was therefore at home. Upon seeing and hearing us approach, she opened the front door and says, "so he's fired you at last, eh?" and we all sat down with the van driver and had a nice cup of tea!

Iris on a beach near Sligo, Ireland circa 1967.

Iris at the Twelve Bens, Connemara. Norton model
19, 600cc with Symes' Jet 80 sidecar - 1967.

Chapter Nine

Vauxhall Motors – 1968 to 1979

I had about three months unemployment before I started work at Vauxhall Motors, Luton and during this time, I designed a 125cc twin cylinder sleeve valve two stroke. It came about thus, for some time, in the motorcycling press, a controversy had been raging about a British world beater to take on the Japanese in the lightweight classes i.e. 125/250cc and people like Phil Vincent and Bernard Hooper, Joe Ehrlich, etc, were putting up proposals. (All of this can be read in my back copies of the motorcycling magazines.) A lottery was even started and approximately £7,000 was initially collected which went, I think, to Velocette and BRM where Peter Burtham, the designer of the V 16 BRM racing car engine, was going to design the new 'World Beater'.

The 'young' Collett now thought he should contribute, and I wrote an article, published in the March 1968 edition of the *Motorcycle Sport*, advocating a 125cc sleeve valve two stroke, with under piston primary compression and four stroke type lubrication for the bottom end. Several people wrote to me expressing various viewpoints and a Major Treen, latterly of the REME invited me down to Sussex to view his 'World Beater'. I went down, combining it with a visit to my parents and the in-laws. He showed me his comprehensive workshop and the engine which he called Excalibur, it was a four-cylinder radial, uniflow scavenge two stroke with inverted pistons exhausting over the skirts. It now resides in Sammy Millar's museum at New Milton, Sussex.

I now started at Vauxhalls and had a 27-mile journey each way on the Norton which I found a bit much after my short trip to Bletchley. Two incidents, on the way to work spring to mind:

1. I was coming down the A6 from the north going to work, approaching the Clophill/Maulden crossroads, when a car suddenly shot out from the Clophill side and on seeing me, stopped in the middle of the road. There was no chance to avoid it, so I laid the bike down on its side and slithered into the car, the wheels of the bike taking the impact, which shot the machine upright throwing me over the bonnet and further down the road. He turned out to be a Sergeant in the USAF at Chicksands and had only just arrived from the USA. Fortunately, nothing was broken on both man or machine and I was able to continue on my way! If I had hit him head on it would certainly have smashed up the front end of the bike and me. I like to think that my rapid reactions were helped by vintage racing! He did later to come round to our house to see if all was well and offered to pay for any damage.

2. Riding to work one day in winter with snow and ice on the roads, I was going gingerly through Box End, Kempston when the bike suddenly snapped away from me And I sat down heavily on the road, on my lumbar vertebrae, I managed to get the bike (Norton in both cases, sold later to John Martin at Hunting's) home, but had to have a week off work lying on my back to recover, luckily nothing was broken and I have not had any further back trouble since!

However, our holiday this year consisted of a fortnight in the Isle of Man with John, Lyn and the Symes' to take in the TT. Iris, Miles (only eight months old) and I motored up to Liverpool in the Hillman van and left it there, meeting up with the Lanes to cross to Douglas where we all got aboard the Norgan Superior and sidecar (see photos) to cross the island to our lodgings at Peel, where we met the Symes' who were lodging down in the town. We were lucky this time, the two stroke Yamaha, 4 cylinder, two strokes were more than a match for the Honda's, winning both the 125 and 250cc races, Bill Ivy breaking the 100-mph lap for 125cc machine! Incredible.

Three amusing, possibly tragic incidents occurred with the Symes'
kids! not necessarily in this order: -

1. During the watching of the TT races, whilst Richard and
   Ann were distracted, Michael and Peter started a game of
   football, initially with a ball but later a man's camera,
   admittedly inside its leather case. He, of course, was furious
   and on approaching Richard was informed that 'boys will
   be boys', reaching for his chequebook; 'just let him know
   how much it was going to cost!' Unbelievable!
2. Our hotel was out of town on a steep hill and one day whilst
   visiting us, Richard was as usual rambling on, when Anne
   gave a shout as the car was seen to be rolling down the hill
   towards Peel with the two boys inside! Richard legged it
   down the road and managed to get inside it, the lads had let
   the handbrake off – it could have killed someone!
3. We heard later that the hotel manager had been awakened
   in the early hours of the morning to find the Symes' kids
   raiding the kitchens for food, to be informed by the parents,
   "Oh well, they get hungry at night and always get up and
   help themselves at home!" say no more, they were a bloody
   nuisance – complete lack of discipline!

Meanwhile, we toured the Isle of Man on John's bike with Iris and
Miles in the sidecar; me on the seat at the back of the sidecar, and
Lyn on the pillion seat! we went all over the South Burrell on
tracks and grass roads, great fun!

It was the September of this year 1968, that my mother died. She
had been ill for some time, eventually ending up in Saint Helier
hospital, Carshalton, where she remained for four months – they
kept you in until you were more or less cured in those days. A few
months later Mr Fenney also died, leaving Martin and his mother
to soldier on their own, as were Dad and Phyllis. About this time
Clive Woodfine moved from AGI to live in Malden and work at
Hunting's in Ampthill, where Iris and I would visit occasionally
from Turvey.

One evening when I was returning from Vauxhall in the van there was a blizzard and the A6 north of Luton was blocked with traffic moving at a snail's pace. We crept down the Barton cutting, through Barton, slipping and sliding on the treacherous conditions, it would have been difficult to have stayed upright on the bike! However, when we got to the Clophill/Maulden crossroads (no roundabout there then) I decided to turn left and make my way through Ampthill – my bad decision. I got as far as the Waters End turn and was defeated by the hill, having to abandon the vehicle there and walk! By now, it was about 8:00 o'clock it had taken three hours to cover some ten miles, the van was left in a snow drift! I walked a further mile and arrived at Clive's house, who phoned Iris, and put me up for the night. The following morning was a Saturday, bright and fine, Clive and I went back to the van and persuaded a local farmer to tow it out of the snow drift – and I proceeded on my way.

Nine months later came the first moon landing and Iris and I watched it on Williams' TV next door, in the early hours of the morning (greatest thing that has happened in my lifetime – it has been a privilege to have lived at this time, when more has been learned in a few decades about space than the whole of the previous centuries) marvellous!! Following this, we moved to Flitton in the autumn of 1969, this was made possible by the death of my mother, who left equal shares to Dad, Phyllis and myself in her will, having had money in her own right. I came upon the old White Horse whilst returning to Turvey, after visiting Mick Bodies' house at Brook Lane, Flitton. The purchase price was £8,250, so I kept my transferable mortgage as it was and paid the difference with my inheritance money.

We brought most of our furniture from Turvey by the Hillman van and simply plonked it down in the house as the previous tenants had left all the carpets, curtains, fittings, etc, so that within half a day it looked as though we had always lived here! Within a week or two, we were off on holiday with the Lanes to Newquay in Cornwall to meet Bob Durrant, Margaret and John's brother,

Mike with his family and wife and two girls (see photograph). John and Lyn came up to Flitton in his first Rover car FLY19 with a great wooden homemade trailer with a large tin wardrobe, bought from Woolwich Arsenal sale, with lorry tyres cut in half for mud guards, it looks terrible. When we arrived at Durrant's posh campsites, he at first refused to acknowledge us! However, eventually we got in and put our scrappy tents up amidst all these posh caravans, mobile homes, tents etc, we had a good time, weather reasonable and Miles walked quite a long way along the cliff paths around Newquay, he was only about 11 months old! It turns out later that Durrant's campsite wasn't so posh after all – it was at the end of an RAF aerodrome runway.

1969 was the first year I raced Richards Triumph 500cc and we started at Cadwell Park, he trailered it up from Goodleigh, Devon where he was now living, having been promoted to works manager of a new branch of the firm he had worked for in Kingston, Surrey. It was fast but fragile and I retired in both races, the Magneto came loose, clutch overheated due to having a standard road chain case and other bits fell off, not wire locked, and he had fitted 3134 cams and twin Carburettor's alright for top speed, but difficult starting and little mid-range 'grunt', also the handlebars were too wide for me, I liked narrow bars so you could tuck in. However, unlike Jerry he was prepared to listen; he had in all faith set the bike up for the road, but racing is different and whereas you can ride many miles on a road bike without trouble, once on the track, being thrashed, is a totally different story! So, when he turned up again later at Brands Hatch with a bronze cylinder head, 'milder' cams and one carburettor, we streaked off the line and finished second! Followed by a third place at Cadwell's end of season race, with the Scott/Norton filling in the gaps. Incidentally, I never fell off the Scott/Norton in a race and I put this down to the very low centre of gravity with the horizontal engine layout.

It was during the visit of the Lanes to Flitton for the Durrant holiday, near Newquay Devon that we met Jack in the autumn of 1969. It happened thus, we had been for a walk on Flitton Moor when, on coming back up Brook Lane, John suddenly exclaimed "Look, there's an MK 8 Velo," and immediately disappeared up the front drive of this cottage, with Lyn shouting, "Come back, it's private property," from then on Jack's fate was sealed! He lived there with his old mother and father having bought them with him from Ash Road, Luton in 1963, his father was a gentle old man who spent most of his time in front of the fire reading western novels, he died in about 1972 whereas his mother lived to 90 odd and died sometime in the 80s, after moving to an old folk's home at Steppingley.

He introduced us to lots of his own interesting friends like: – Ken Baldwin who had flown Dakota's in the Berlin airlift in 1946 and who now lived in Grassington, Yorkshire. Reggie Turner, and old motorcycling pal from his teenage days in Luton, Dennis Murphy, who regularly visited him to go shopping at Tesco's and was also an old school chum who flew Spitfires etc during the war; Ted Stone from Eaton Bray, who had spent many years in Africa and invented two types of ball engine (see photos) and had his own workshop etc and went with us to several lectures at Cranfield and Silsoe universities, on engine design, one by Gordon Blair of QEB racing fame, plus of course, all of his large family including his brother George he worked for Rolls Royce and sister Rene Aspel who lived in Luton, where most of her large family resided at that time.

Our second holiday this year was a trip with Miles to a B&B farmhouse in the Brendon Hills, Somerset. We went in the van and stayed a week, during which time we explored the area with frequent visits to the north coast seaside resorts. Originally it had been a mine of some sort with an incline railway running down to Watchet seaport; the remains were still visible and interesting. It closed eventually after a series of accidents when the trucks came loose and ran downhill killing several workmen, etc. Whilst we were there, one morning at about 8:30am before I was up Jack arrived on his Rudge, having ridden down from Flitton for the day and returned the same evening. A good 400-mile round trip, to be sure!

Jack immediately started to come to our vintage racing meets, bringing Iris and Miles in his Austin A40 car, with me having gone ahead with the bike in the van for scrutineering and early morning practice. This went on until Sarah was born on 26 March 1972 and I finally retired on 16 September 1973. I don't remember taking Sarah to any of the last meetings which were completed using Richard's Triumph, The Scott/Norton was virtually retired as I was now building up a workshop in the blacksmith barn to start manufacturing my own designed sleeve valve two stroke engines! Whilst all this was going on, I was also rehashing the Triumph two stroke for its third and final phase.

I purchased a 1947 500cc Triumph speed twin motorcycle with telescopic front forks and solid rear end, into which I grafted the now single blower engine (I have no photos of this phase but there is a drawing of Jack's showing the engine installation) And started to use it on the roads to go to work among other things! I used it for several months even taking it to Woburn Abbey with Jack, Iris etc and the Lanes to a motor/motorcycle gymkhana where the bike was entered for the most original 'special' which it should have won with ease.

Unfortunately, the judge was Bertie Goodman of Velocettes and as Headley Cox points out in his book *From the Race Shop Floor* he was a bit of a twit; and gave the award to a man who had put a Volkswagen air cooled flat 4 engine in some frame or other – routine! So, I continued using it for work as daily transport until one day, after leaving work I notice a group of the lads from the office standing at the bus stop in Park Street, Luton. So, I decided to demonstrate its superb acceleration! There was a shattering bang and bits of engine splattered my Wellington boots, the cylinder barrels had sheared off around the transfer ports. They of course, fell about with laughter as I had the indignity of pushing it back to the bike sheds and joining them in the bus queue! I retrieved it the following day with the van.

Its final demise, however, was to be stolen by Jazz Parkes, a young General Motors student who was studying for his qualifications to become an engineer and had shown a great interest in the project, coming on occasions with us to some of the vintage racing circuits. So, as I was tied up with other things, I offered to lend it to him to rebuild and continue with the development; such is my generous and helpful nature! I nearly lost the Scott/Norton in the same way but was warned in time by Dave Leqoc, Whereupon Jack and I went round to this chap's house and retrieved it in bits; he was going to do it up and race it against Chris Williams, who was the reigning vintage champion of that time. Eventually I took it to Roy Sherwood who rebuilt it back to its present form (an article was done about the restoration in the *Motorcycle Sport* of which I have a copy).

However, I digress, back to the Triumph. After a year or so he left Vauxhall, but I wasn't worried as I knew where he lived in Tebworth, I had visited on several occasions to view progress. Eventually however, having not heard from him, I paid a visit – no answer at the front door then his neighbour informed me that he had moved, leaving no forwarding address – *fait accompli*. I find this rather 'galling' for obvious reasons, plus the fact that the Shorroks Supercharger was worth considerable money, even if the bike wasn't, but I would have liked it back to complete the set of specials I have built during my lifetime. P.S. I still have the logbook!

Meanwhile, I was designing 125cc sleeve valve two strokes with under piston transfer delivery, the EO as shown on Jack's first illustration that he did for me. In order to understand the manufacturing assembly of it I asked Fred, who I had heard was good at illustration, to prepare some sketches (which I still have) and he asked if he might come and help with the project and he has been coming ever since!

Now starts the period of intense activity initially with the building up of the workshop facilities; we paid a visit to the in-laws at Beddington and on the way back called in at a firm that was selling off their assets and presumably going into liquidation.

I purchased a large electric motor and its line shafting, pulleys and plumber blocks etc and this formed the basis of our workshop at Flitton for many years. (See photographs). It drove a horizontal mill, obtained from Rex Boyer and a lathe from somewhere? We used Jack's hand cranked drilling machine mounted on the bench that had come from Turvey and the South Bend lathe arrived early from somewhere near Hatfield, that was it for quite a while before the shaper, drill press and fly press arrived.

We purchased quite a lot of the small tools and measuring equipment from Chiswick supplies at Melbourne, Cambridgeshire. they are still in business I believe, but a breakthrough arrived in the shape of Bagshaw's of Dunstable who were closing down. It is called 'asset stripping' and it's one of the plagues, like 'developers' who are trying to and succeeding in covering the country in concrete! We went along and purchased at knockdown prices such things as a dividing head with all its accessories, rotary table, Vernier's, micrometres, machine vices, etc all of which set us up to a good start. The metal for the engine, we obtained from a Sam Saunders who was the workshop foreman of a firm called Hudson's, close to Vauxhall's. So, I could order it at lunchtime and pick it up on my way home in the evening, very convenient and he was a very useful source of information with his vast practical knowledge.

Also, at about this time, Steve Linsdell came into our orbit, I had originally encountered him on my way home from work on the Norton, when we would have a bit of a 'dust up' from Barton to Pulloxhill around the lanes but I didn't know who he was then, except that for a young lad he could ride well, little did I realise! He joined us on Thursday evenings for a drink at 'The Bell' pub in Greenfield with Jack, after we had finished working on the engine, the pub is still going on now but with slightly different people!

One amusing incident springs to mind; on this occasion we had finished work as usual at around 9:00 pm and I, with Fred and Jack behind me, entered the kitchen to wash our hands before retiring to the pub; to find Iris hacking at a chicken with a hatchet, presumably attempting to prepare it for tomorrow's lunch! Whereupon seeing us, she started to get even more excited and started shouting abuse; so, I naively tried to calm her down, suggesting if she didn't, she might cut her fingers off; to no avail it only seemed to aggravate her more! I turned round to appeal for support from the lads only to find that they had 'done a runner' and were no longer behind me. Traitors! I felt like I had just gone 'over the top' without support. Upon arrival home, a couple of hours later, I thought this incident closed but no, she was still infuriated and in the ensuing argument, snatched a mirror off the dresser and hurled it at me! Luckily, my quick reactions saved the day but not the mirror! She would periodically have these rages and on another occasion, threw a potted plant complete with earth across the living room with similar results! But most of the time she was reasonable, I think she took after her father. He used to upset his friends at his local pub – the Harvest Home in Beddington.

Now the first 125CC engine, numbered E0, was a vertical twin cylinder arrangement and it soon became obvious that I had bitten off more than we could chew! It was proving difficult to make and we hadn't tackled the most difficult bits yet.

So, we stopped after eight months work, on 3 September 1973. We know this accurately because we kept a log of every engine we had produced! See article in *Motorcycle* 9 December 1972 by Vic Willoughby again, with Jack's first EO drawing, beautiful piece of work! It makes interesting reading after all this time especially the bit about going into production, I don't think I ever said that! Willoughby must have mistaken me, at no time on any of the projects have I ever thought of production, in fact, just the opposite, I am all for leaping on forward to the next challenge, so much so that I've often had trouble dragging Fred with me! He preferred a slower and more methodical pace. A single cylinder version of this engine was previously described as 'The British World Beater' in 1968, in *Motorcycle Sport*.

It was very difficult decision, especially as Fred and Jack had done a great deal of work on it and I alone bore the responsibility for it. However, it was the right decision I'm sure, the second 125 EI had less parts, all easier to make and went on later to a successful conclusion at Cranfield University as you will see. So, we beavered on for three years until it was finished on 28 May 1976 after 1,259 hours of work. Then we set too and started to make a Dynamometer, we like to do things the hard way! To test it on. It was built on a trestle (see photos) with the engine one end and a set of disc brakes the other, with a reaction gearbox, gimbled on bearings to an arm operating on a spring balance. A large 2HP electric motor stood underneath with a connecting belt and jockey pulley for starting and declutching.

On the first run, the sleeve drive trunnion bearing seized at about 5,000 rpm, it was a plain bearing and was replaced with needle rollers giving no further problems up to 12,000 rpm on occasions! We had various troubles with both engine and rig, but the disc brakes proved unsatisfactory due to unstable performance and were replaced with a Telmar Retarder, which is an Eddy-current electric transmission breaking system fitted normally behind the gearbox of coaches and heavy lorries etc, to assist the normal brakes when descending long steep hills etc. It was supplied by Stu Stringer courtesy of Vauxhall Motors, on loan.

This proved very successful using a 24-volt system controlled by a rheostat to apply the load, we recorded 18 brake horsepower at 11,000 rpm, after many trials and tribulations (see the logbook). All this took us through the 1970s to 29 November 1980 when the last run was recorded with this set up. Several incidents occurred during these trials, the most serious which could have had unfortunate repercussions was when Fred was blown up! It happened like this, due to the noise pollution etc we had fitted a 40-gallon oil drum to the rig which totally enclosed the exhaust expansion chamber etc and allowed the spent gases to exhaust up the Old Forge chimney via a long-segmented pipe and if the wind was in the right direction, into Brooks back door. Thus, in addition to not being able to hear his television, he now wasn't able to see it because of electrical interference caused by the unsuppressed high-tension ignition and now he had exhaust fumes to contend with! Rosemary, his wife used to come round to try and complain and would be seen at the barn door jumping up and down with rage, but we couldn't hear her with our ear defenders on! This oil drum had a frangible diaphragm made of hardboard in one end, as a safety valve should an explosion happen to take place.

We had just finished a run when Fred, for some reason, stepped in front of this diaphragm at the same time as I switched off the ignition. The result was a huge bang and a sheet of flame that enveloped him! There was a moment of silence and then Fred was staggering around the workshop saying, "what happened, what happened?" I rushed over and started beating out the flames on his smock etc, just as Iris appeared, took one look at him and said, "I'll go and make a cup of tea." We got him sat down, his eyebrows and moustache were half gone, and he had been peppered with shot from the remnants of the hardwood diaphragm. By this time, several people had arrived, attracted by the bang, including the Austins from the Vicarage over the road. After about 30 minutes he had recovered sufficiently to joke about it, likening himself to a human cannonball, as sometimes seen in circuses! When he went home that evening, he received little sympathy from his wife, her remark was, "if you will go around and play with that Collett chap, what do you expect?"

She may have been right! Cruel. In retrospect, what had happened was after the run, the engine being a wasteful two stroke had filled the drum with a mixture of burnt and unburnt gases that had not yet been pumped up the flu and into Brooks kitchen. The engine contact breakers must have been closed and therefore when I switched off, a high-tension spark occurred at the plug, travelled down the open exhaust parts, with inevitable results!

On another occasion, one Saturday afternoon, we later discovered Sarah Austin was being married; they had difficulty in hearing the service and smoke can be seen drifting across some of the wedding photos! We had some cold winters during this time and one Thursday night we were both hard at work when suddenly the barn filled up with acrid smoke; the big line shaft electric motor had caught fire, I rushed over and turned off the switches, then followed Fred outdoors. we never found out exactly what caused it, but later when switched on again, it ran perfectly and continued in this vein for another two years before repeating the process, this time it did not restart.

The third time we tried to burn the workshop down was on another Saturday afternoon session when our paraffin stove, donated to the project by Fred's Mother, ran out of fuel, as we had no more on the premises I went round to Jacks to scrounge some; he produced a can with paraffin written on it, and we filled the stove. Upon ignition, the whole thing caught fire instantly; it was petrol!! I grabbed it and rushed out of the, fortunately open, doors and threw it on the drive to burn itself out, returning to help Fred douse the remaining fire which had flowed out over the floor! I should have checked knowing Jack!

On several occasions we would have visits from Symes he would stand over Fred on the South Bend lathe, incessantly talking as usual, whilst putting smoke from his pipe over him! Another visitor was Dan, Phyllis's husband, he would stand wittering to Fred, wasting his time explaining in detail how he made Eccles cakes! We did some useful work however, for during this time Dutch Elm disease had caused the felling of all the lovely trees on the Avenue leading from Silsoe to Flitton and the council had left them in convenient lengths for anybody to pick up, as I did on my way home from work each evening in our shared Dormobile van. This, Jack and I had bought from Stu Stringer for £100 pounds i.e. £50 each, the idea being that he and Iris would share it during the day, and I of course, went to work on the Norton.

I hired a chainsaw and Fred and myself spent one or two Saturday afternoons sawing them up, some he took away for his mother,

Jack had some and the rest were put in the cellar for our use, having reordered the front small room, originally called the 'snug' bar when it was a pub in 1967 and run by Mrs Catlyn; to improve the fireplace etc so that we could have open fires.

Early in the 70s, Iris's parents were brought up from Beddington to stay and my father came for Christmas bringing Uncle John, my mother's brother who was in care, administered by my father for many years, as he was, to use a modern term 'low grade' and had been since birth, I eventually took over the responsibility after my father's death in 1993 and I remember my mother saying that they hoped he would die before them, so that the responsibility would not have to go onto a further generation. But as in all things, 'the best laid plans of mice and men!'

However, I did not have to do it for very long, only a couple of years if I remember correctly. He was at this time in a home in Herne Bay, North Kent and we visited him once or twice when staying with the Lanes. When he died his funeral took place in Herne Bay Crematorium and it was attended by John and Rosemary, Phyllis, Jill and I and a couple of the nursing home staff. He had little effects left over to show for his long life, just a few trinkets in a bag which were handed to me at the funeral, what a dreadful shame; some people get a 'raw deal' from life!

Getting back to comedy; Iris and I went to a Vauxhall dance evening where Fred and Jeanette, Chris Plumber and wife Valerie, plus many others of our office staff were there. Things were going well until about 11:00pm when Iris started to get 'shirty' and demanded to go home but I was enjoying it jiving around on the dance floor, when all of a sudden, my overcoat came whistling through the air and landed in the middle of the dance floor! How embarrassing is that! So, we went home. At work, I joined my first union the 'Draftsman's and Allied Technicians Association' (DATA), which at that time was for technical stuff only but later on started to embrace firemen, administrative staff, even cleaners, so its whole identity was lost; they just wanted numbers to hit the employers with.

I did a year as office representative, a thankless task, you get little support from the members; just gripes and disapproval from the management, plus I am sure that it affects your salary and promotional chances. It's a job very few people want to do, so the same old people end up doing it year after year and I felt that if, as members, we were to receive the benefits i.e. pay rises, shorter working hours, holidays with pay etc, we should put something back! Also, at this time I started to work with Derek Keep and joined his newly formed 'UFO Club', which eventually went on for 35 years; astonishing! Although there was not much UFO about it later on, more of a social club. Around this time, Mick Boddy was into hand gliding and had built himself an early style arrow shaped, three pole, canvas covered glider. Somehow or another the UFO Club got involved and we all retired to the Warden Hills, North of Luton for a demonstration and possible flight!

The demonstration was successfully carried out and then Mick asked if anyone else would like to have a go. I think I was the first to try and after some explanation ran down the slope grasping the control bar firmly with both hands; I was amazed how quickly I was jerked into the air, to a height of about ten feet I suppose, so trying to control it I pushed forward on the bar and almost immediately nosedived into the ground! Landing on my feet but overbalancing to end up face first in a small bush with the machine draped all over me. I was unhurt but embarrassed, needless to say and had travelled about 30 yards or so. Jack also had a go and more or less repeated my performance, in spite of being a Spitfire pilot! I think a couple of others, Geoff Perry, springs to mind, had a go with similar results before Mick called it a day, to save his hand glider from being wrecked. Very interesting we needed more practise and the ground was not very suitable being covered with potholes and extraneous vegetation!

Derek Keep at this time was Vauxhall's stress man and I learned a lot from him, for which I am very grateful! He was very kind and would help anybody, even though he was a member of the

National Front and supported Enoch Powell, I still keep in touch with him. About this time 1971, Phyllis decided to get married to Dan from Wellingborough, whom she had met at some do up in London where she was working as a typist at Somerset House.

The wedding was to say the least, a riot; it was held in Wallington at our 'local' parish church and the reception was in an adjacent hall where all the food was laid out buffet style. What we hadn't realised was that Dan was a member of a family 13 children strong with all of their offspring etc. they just descended, like locusts, on the grub etc, whilst we, that is Dad, Iris, Uncle Arthur, Aunt Dorothy, me and friends struggled to find anything! We later retired back to Aunt Dorothy's house to recover and she laid on high tea. Dan was renting a house in Wellingborough and it was to this that they returned; later with the sale of the family silver that had been left by my mother (it was decreed that it should descend through the female line, hence I was excluded from being a beneficiary) and her one third share from mother's estate they bought a terraced property in Alexandra Road Wellingborough.

Some months after this my father had a second breakdown; he had had a previous one in the mid-1950s, so severe that he was sent to Fairdene Mental Hospital for treatment. – Electro Convulsive Therapy (ECT), from which he made a good recovery although he was off work for about nine months. The first I knew about it was a phone call from Aunt Dorothy/Arthur with whom he used to play golf, asking if I knew where Dad was. It appeared that he had not been to the Golf Club for a week or two and a visit to his house proved to be of no use.

So, I went down on my motorcycle and made inquiries; Mr Muston, the next-door neighbour thought he was in Belmont Hospital in Sutton and so it duly turned out to be. I bought him back home and returned to Flitton in his car with him, whereupon he stayed a week or so to recover before returning to Wallington again. This time, fortunately he stayed and eventually made a full recovery by occupying himself with golf, bowls, whist drives, joining 'Probus', and an archaeological society. It is a great shock to lose a partner after so many years and although he seemed to get over my mother's death, he had Phyliss with him for support but when she went, that prop was withdrawn! It is a great problem now and the authorities are just beginning to realise as people are living longer, so help lines are being set up. Loneliness is a great problem and I often think I should have done more for my father by having him to stay more often.

A holiday in Norfolk on a heath near Sheringham, in Pete Sugar's caravan took place around this time with Iris, Miles and me. The weather was wet most of the time it seemed, but Miles didn't complain, he spent most of his time playing in the puddles outside the caravan, I think he preferred that than going to the seaside!

Several articles appeared in the motorcycling press during this time. Using Jacks illustrations, describing progress and explaining the 'advantages' of sleeve valves? I also gave a lecture to the students at the agricultural college at Silsoe on Vintage Racing, the Triumph two stroke and sleeve valve progress. It was well received, and I hope stimulated some students to do likewise, as I have met many who seemed to think that unless you have a vast organisation behind you with sophisticated equipment, plus plenty of money and a large team of experts, it is impossible to produce anything worthwhile. Not so, it is a myth generated by the university education system, which is run by academics, most of whom have probably never made anything in their lives, and remember, an engineer is, to quote Neville Shute, 'A man who can do for five shillings what any person can do for a pound!'

1976 was the year of the hot summer that everybody talks about, but it didn't last beyond August. I know because I kept arriving at work on my Norton dressed as always, in my dispatch riders, coat and Wellington boots etc, to ribald laughter from the underdressed idiots, most of whom didn't realise that when falling off a motorcycle one requires protection! However, I had the last laugh as I used to say to them 'it won't last' and of course one day it came to an end. It started raining heavily during the afternoon and come five o'clock I found a large crowd of people at the exit doors, wondering how they were going to get out without getting soaked, you can imagine what I said to them! See the book *The great drought of 1976* by Evelyn Cox, of which I have a copy.

It was also the year we built the wall on the left-hand side of the house, using second-hand bricks and a church door we obtained from Rochester church on a visit to the Lane's, in fact we bought two, the second one came up to 'Stanton' and is now installed next door at Number 51.

The Lanes had a big do this year and I met Ray and Jill really for the first time (there are photos of us all dancing around the campfire in the early morning). The noise was terrific, Glenn Miller was being played over loudspeakers set up in the trees when all of a sudden police officers appeared through the woods and surrounded us. it transpired that the neighbours had phoned them thinking that it was a drug induced party, the only drugs were alcohol and jazz!

We had a holiday in a little cottage at Henstead Hall near to Southwold in Suffolk, which was recommended by Barbara who lived further down Flitton High Street. Barbara said the accommodation and location were good but the food... Well, it was reasonable, but they didn't give you enough! On our way there, we called in to see my old army pal, Graham Vallor, previously mentioned. what a shock! He was dead, had taken his own life. His mother explained that he had 'got in with a bad lot'. Now, I don't know what 'lot' it was, whether religious cult or drugs etc? I didn't like to ask but it shows how careful you have to be when choosing your friends. He left a wife and baby daughter whom they had called Collette. they were all living together as Mrs Vallor's husband had passed on also. I have no idea where they are now! Should have kept in touch.

The Three Torrs run started circa 1974 and went on intermittently until about 2000, around 25 years! It was set up initially by Richard Symes and Ron Ley who were both members of the Devon section vintage club and the idea was to start at Torrington and ride down into Cornwall i.e. near Camelford and climb Brown Willy, the highest point on Bodmin Moor, then proceed via Oakhampton to Dartmoor and climb High Willhays followed by a further run northwards to Dunkery Beacon on Exmoor. This little jaunt started at 9:00am and finished at 9:00pm involving about 170 miles of motorcycling and 10 miles of rough walking, quite a strenuous day! The first riders were Jack and I from Bedfordshire and John Lane and Derek Clansy from Kent. We used to meet up at The George at Frome for a pint and a snack before continuing to Goodleigh while we stayed with Richard and Ann Symes.

After a couple of rides or so, Steve Linsdell joined us on an Indian Enfield 350cc Bullet; he had just given up working for his father at Flitwick nurseries repairing lawn mowers and other gardening tools and had struck out on his own, initially in his father's premises, later Flitwick motorcycles in Station road, Flitwick. When he first started, he was an agent for Enfield's and MZ, later moving to Yamaha. As time went on and after Richards death in 1989 others like Miles, Shed, Nick, Peter Wheeler, Heasman and Fernando came and the whole show moved to Geoff Brown's bungalow also in Goodleigh, with the breakfast stop now at Clive Heasman's house in Devises, Wilts. (see photograph.)

Some of the incidents during these years may be of interest, surprisingly we had no major problems of life and limb. Early on when we stayed with Richard, we found out from the locals at the New Inn pub that he was not particularly popular. One, by being a 'foreigner' and two by his somewhat bombastic attitude, and the fact that he insisted on playing his record player at full volume with the French windows of his house fully open!

We were returning from an early Three Tors run and separated somewhere near Shepton Mallet, Jack and Steve going on to

Bedfordshire, whilst I decided to accompany John and Derek over the Marlborough Downs to pay a visit to see John's parents who had by now moved from Beddington to Hindon in a small village some 10 miles West of Salisbury, Wilts. We were some twenty miles away when John's bike stopped, investigation showed that the valves were not operating, they had open valve gear in 1927. It turned out that the lower-level gear Woodruffe key had sheared on the crankshaft. What to do? We were in the middle of nowhere! On the downs, well what we did in the end was to cut a washer of approximately the right size, in half, file it roughly to shape using a pair of mole wrenches as a vice! It worked and got him back to Walderslade as well, not without a few problems. As he entered Reigate, I believe the police stopped him for having no lights; we had lost so much time with the repair but incredibly, instead of detaining him and booking him, they said, 'follow us and we will get you home.' Can you believe it, so he had a police escort with blue flashing lights all the way from Reigate to Walderslade, a distance of 30 miles at the taxpayers' expense!

Jack was the victim of the next incident, we were all returning home and had got as far as Shepton Mallet when I, who was leading on my MZ saw a large stone right in front of me. I swerved aside but Jack he was right behind me hit it and fortunately managed to stay on. It put a large flat section in his front wheel rim but incredibly did not puncture the inner tube, so he was obliged to ride home in a series of bumps! He later had to have the Rudge front wheel rebuilt but maintained the 21-inch specification.

On another occasion, I re-established contact with an old De Havilland friend, Keith Bramwell who had been my old boss when we worked together on the Constant Speed Unit (CSU), section of DH Propellers Stevenage, it happened thus: we were staying with Syme's and on the Saturday before the run, I had had a skinful of rough cider at the local pub at lunchtime and was still feeling somewhat second-hand when a trip out for an evening meal was due.

This was to take place in Fremington, a village outside Barnstaple. During the meal of which I could only face the soup course, a woman came over from an adjacent table and addressed me directly: 'are you Bob Collett?' she said.

'I think so,' I said.

'Well, I am Maureen Bramwell and Keith is sitting over there!'

Amazing! They had decided that evening to go to the same pub with some friends for a meal – they say, 'ships that pass in the night'! The last time I saw Keith was at De Havilland's, when I left during my nervous breakdown and he asked me later whether he had contributed to it, perhaps putting too much work pressure on me. I assured him this was not the case. He lived further up the road in the old schoolhouse (see photos) And I still keep in contact with him at Christmas.

As I write this diary (January 2015) I have just received a card from Maureen, informing me of Keith's death, from lung cancer! Which is slightly strange as he never smoked or drank particularly, as far as I know. He had been ill for some six months, but the doctors had missed it, only diagnosing it properly too late! He was a nice chap, a very clever engineer, but unfortunately in order to earn a big salary in this country, he had had to switch to management! When I informed him of Jack's death last March 2014, he said, 'he was a lovely, lovely man' They were both artists in their own ways, Keith had a studio in which he made sculptures.

Weather wise, the worst Three Torr's we had was in 1989, the year Sarah went in the sidecar. I had just purchased this immaculate MZ sidecar from Mick Boddy for £250 plus another MZ in bits and various MZ oddments, including a spare engine, which Jack shared with me, the whole conglomerate spent years residing in Jack's barn/garage. Sarah and I travelled down on the Friday, going directly to Geoff Brown's house, where we met the others. This was the first year at Geoff's, as Richard had died earlier this year.

The weather was wet from the word go, we were soaked by the time we got to the start at Torrington and headed South West to Cornwall, Bodmin Moor and Brown Willy. We abandoned the idea of climbing the mountain, and all set off to Dartmoor, there were many breakdowns – water in the electrics etc, and we all ended up in a cafe somewhere on the way to Oakhampton. It was decided by unanimous consent to abandon the run and head for home, by this time Sarah was sitting in a bath with water filling the bottom of the sidecar body. Luckily all our machines kept going and as soon as we arrived back at Geoff's, the sky cleared, and it was a beautiful evening! However, ill luck was not finished with us yet, on the way home the MZ failed at Frome, and Sarah went back home on the rear mudguard of Jack's BMW leaving me to find a bed and breakfast and arrange for repairs be carried out. I contacted De Fazzio's of Devises and they sent out a mechanic, but he failed to fix it. It later turned out to be a dud ignition coil! I went home on the train and later John Urwin and I went back in his Jaguar and trailer and bought it back, what a pantomime!

The first Three Tors Steve did on his 'Black Pig'; this was a 1200cc Royal Enfield 1938 V-twin side valve machine. It caused a great deal of trouble on the run down with one problem after another, until eventually the fuel tank split and required welding. A garage was approached but they would not consider it for obvious reasons. After money had changed hands, they agreed to let Steve do it, so we spent some time flicking lighted matches into the filler hole, watching a sheet of flame leap out until he judged it safe!

In 1995 I went on the four-stroke sleeve valve bike with Ian, Jill's youngest brother; it was the first of four Three Tors runs I did on it (see logbook). It performed reasonably, recording a fuel consumption of 85 to 90 miles per gallon, same as John's CSI Norton! Both better than anyone else but we don't talk about the oil consumption of three litres for 750 miles, surpassed only by Steve's Black Pig!

Next year, I had problems with ignition timing and had to reset it once on the A5 going down to Flitton and again at

Stow-on-the-Wold, when it seized on the way back. On a later occasion, I shot through an open farm gate into a field when descending a steep hill somewhere near Porlock, the front wheel under heavy breaking had come up between the two front exhaust pipes and I couldn't turn it to go round the bend at the bottom! And finally we followed a farm tractor for several miles with bales of cotton falling repeatedly onto the road and bouncing dangerously close to our bikes but the man in charge didn't seem to notice, until eventually Shed in a fit of bravery overtook and waved him down; whether or not he went back to collect them, we don't know.

In 1999 Jill and I went down, not on a Three Tors but later in August for the total eclipse of the sun, totality being only visible in the UK in Cornwall. We went on Jack's Rudge which he generously gave and made over to me at this time, for I think two reasons:

1. He was finding it difficult to start, a 500cc single cylinder engine takes quite an effort even with a valve lifter.
2. I had always said that this was the best of all his machines i.e. the BMW, MK8 Velocette, Dotty (350cc Cammy Velo), because of its magnificent four valve cylinder head layout, with central sparking plug; in fact, it is almost universally used now on high performance engines! It was a very generous thing to do and I hope we somewhat repaid him later by taking him on all those wonderful holidays around the UK and Ireland etc.

We went down the Fosse Way to Clive Heasman's house, stayed overnight, met John and Derek next morning with Miles on his Kawasaki trails bike who returned home with Nick, after the event, and continued with Clive to Geoff Browns (see photographs). The Bedfordshire mob including Jack did not attend this one because, I don't think Steve could get the time off.

The eclipse was wonderful, and we viewed it from Bodmin Moor and we were all impressed with the way that the cows stopped mooing and the birds all went quiet during the event, which lasted a few minutes. Our return home up the Fosse Way was a wet ride through several thunderstorms but at least we got home okay. Not so John Lane, he had come for the first time ever, not on his trusty CSI, but on his newly rebuilt Norton Dominator which expired some 20 miles short of Walderslade when the pushrods jumped out of their rocker engagements; the rockers had seized. It turned out on later investigation that the oil ways were blocked with grease which had come from the new timing chain that had been fitted! That bike is still sitting in his garage 15 years on, whilst I'm writing this, awaiting a rebuild. What a shame! Further investigation in 2017 when I bought this machine; it would appear that a restrictor was missing in the scavenge oil line, reducing the pressure to the rockers etc, hence starving them of oil.

A further two, Three Tors took place, the last genuine one in 1998 before Jeff and Ann emigrated to Australia! They had lived in Africa previously for many years. A further so called 'Three Tours' in 2002 when we went towards Hunstanton, visited a Victorian pumping station, saw the sea, etc. I went on the E3 which was the first of the 500cc sleeve valve, water-cooled, two strokes – working on the 'Crecy' principle and Steve went on a BSA Sidecar outfit with Peter Wheeler in the chair, Shed went in a car as he had damaged his arm, Jack, John and Derek went as usual on their respective machines (see photographs).

However, to complete our motorcycling exploits in the Westcountry, I must mention a trip taken in the early 70s by Jack and myself who set off in pouring rain, he on the Rudge and me on the Norton. We were both soon very soaked and eventually stopped at a cafe and ordered something and chips as usual, as I went to eat it, waterflooded down my sleeves and the next minute the chips were floating around on the plate! Never mind, I still ate them. Continuing our journey, we were going to stay with the Symes's we got as far as South Molton when the Rudge stopped; water in the Magneto! Well, we found a garage where mechanics were working with rain pouring through cracks in the roof;

incredible, health and safety would have a fit nowadays! Anyway, we fixed the problem it was water running down the HT lead. Only one other incident springs to mind and that is Symes's driving, for some reason Richard decided to show us Devon! In the car and after several hair-raising miles we asked him to stop and let us out, he seemed to think he was the only one on the road! Terrible. I sometimes wonder if his recklessness was in part to cause of his final demise?

Cast your mind back to 1962 when I built the 720cc oversized Norton engine, well it had languished under the bench for some 12 years until Dick Weaver took it off my hands. Dick Weaver was a draughtsman at Vauxhall who lived in Bath, where he told me there are plenty of hills! He was building himself a BSA Special and I suggested the Norton engine would be just the thing, and so it was. He reckoned it just romped up with little effort, as I knew it would. I don't know what happened, as I lost touch with him after leaving Vauxhall in 1979. It was around this time that aircraft manufacturing came to Flitton, a consortium consisting of Mick Boddy, Les Loosely and another draughtsman whose name I can't remember, had decided to build a 'Pietenpol' which was a twin seater, single-engine, high-wing monoplane of French design. It was to be powered by a Chevrolet air cooled, flat 6-cylinder automotive engine of American manufacture. In fact, it was the engine of the dangerous car that Ralph Nader took General Motors to task for, in his famous book *Unsafe at any speed*. They wanted Fred and myself to join them but as I explained we were tied up with the manufacture of sleeve valve engines etc and would not be able to give 100% commitment!

The driving force was Les Loosely and he had chosen the 'Pietenpol' over more complex designs, one of which was being considered by another group at Vauxhall, this aeroplane was to use a Rolls Royce Continental, flat six engine, obviously built for the job but much more expensive! It never got off the ground so to speak, in fact I don't think they ever cut any metal or wood! Whereas the Flitton flyer made rapid progress initially, all being constructed in one of Mick Boddy's barns, he had two at that time, later they were sold off by farmer Catlin. his landlord, for housing development, say no more! Typical of the man, he already had too much money! An engine was purchased, fuselage, port and starboard wings were made as I remember and it all seemed to be going well, there was even talk of an Airfield being purchased. This was an area of land owned by a farmer, Ron Sharp, which lies now alongside the Ampthill bypass in Hollington Basin, adjacent to the Flitton to Malden Road, totally unsuitable really:

A. Because of the bypass, which was not built at that time, and
B. it is too low lying and is now a lake, but it was cheap! Should have made it a seaplane!

Then the third man (who kept rabbits for a hobby) dropped out, Les and Mick kept on for some time, but I think the increasing cost and workload etc., finally put paid to it. A great shame, nevertheless, it was a great effort and if they could have found someone else, who knows? Or perhaps the landlord was already pressing for the barn's destruction. Eventually, it was all sold I believe and hopefully the project was finished by the new owner.

As you may or may not know, I've always been a Jazz fan, none of this modern noise for me, I refuse to call it music, it's rubbish. 99% of it. I'm afraid it all started around 1956 with Bill Haley and something called 'Rock and Roll', later Elvis Presley and The Beatles joined in, at that time it vaguely had a tune! But nowadays, it's mainly a thumping sound accompanied by whining from some overpaid cretins, whether male or female; lyrics and melody have long since vanished! The reason for this, was when the instrumentalist was overshadowed by the vocalist, let's face it almost all of today's pop noise is vocal. One of our draughtsmen, Pete Brown who sat next to me at Vauxhall, used to, in his spare time, play the electronic organ at various functions and he was very good. He said he gave it up because at the end all they wanted us to do was to give a backing to some new vocalist or other! Any resemblance to music is purely coincidental!

Anyway, what I was going to say was that Harry James, 'The Man with the Golden Trumpet' and his orchestra were on tour in this country and were visiting the Luton Odeon for a matinee performance, so a group of us skived off work that afternoon; terrific. They played many of the favourites including hits like 'Carnival of Venice', 'Trumpet Blues' and the 'Flight of the Bumble Bee' etc. I had always been a fan since obtaining several 78 rpm records from Ernie Evans whilst at school. Tracks from the film *The Young Man of Music* (I have a copy on video).

Whilst all this was going on, I was having trouble with the Vauxhall management as usual. I had been pestering them to move me to the Advanced Design Department on the strength of my sleeve valve engine designs, and upon not receiving any cooperation, was insisting on my rights to see the chief engineer who was an American at this time. GM had for some years now, been replacing our British senior management with their own, in all fairness a large portion of ours were pretty useless! After several negative interviews, I arrived on Jarvis's doormat! He was the 'Heindrich Himmler' of all the drawing offices etc and hated by all. He had been sent down from Elsmere Port as he had caused so much trouble up there that somebody cut the brake hoses on his car and he was nearly killed one night when going home. Therefore, for his own and others safety he was advised to seek a transfer; so, the story goes!

To show what an idiot he was, listen to the following story. One day a cartoon appeared on the office notice board, it showed a large German style jackboot descending from the sky, with a Swastika on it, and cowering underneath were a group of terrified draughtsman, you could tell they were draughtsman by the spectacles, pencil's and compasses. When Jarvis saw this, he went ballistic and demanded to know the perpetrator; some say it was Fred, he is very good at cartoons but there were others equally as accomplished. Of course, the more Jarvis went on the funnier it became! And eventually he became known as Jackboot Jarvis! He should of course have just ignored it and kept quiet. Anyway, when I arrived in front of him and put my case forward, he scowled and said, 'if you weren't in the union, I'd sack you!' Typical! I should have expected no less! Something to be said for unions after all, they certainly helped me later when at Hunting's as you will hear; but I stood my ground and reiterated my demand. It had some effect as I had the backing of the office union committee; only on the right to see the chief engineer, you understand, employees' rights.

My next interview was with Mr Vallas, assistant to Mr Steinhausen, the chief engineer, both Americans. The reason our British lot didn't want me to go higher to the Yanks, was I think that I might criticise their leadership or lack of it, and name names! Paranoid. However, he listened to my gripes, asked a few relevant questions mainly about my work with Derek Keep on the stress group, committed himself to nothing, merely saying that he was reorganising various departments and there might be a position for me; a much more intelligent approach, I thought. Eventually, after a few weeks if I remember correctly, Derek and myself were introduced to a Harry Moss from Advanced Design and under his leadership we were to join a new group called 'Computer Applications'.

Vauxhall at this time was not devoid of computer expertise but it was not coordinated, being left to odd individuals like Philip Pratt, Harry Moss, Jim Curtis, other engineers and myself. I had written a program on push rod valve gear, a bit late in the day as most engines were now moving to overhead camshafts (I still have copies of all the programs I wrote).

All the programs written at this time, circa 1968, where on 'ticker tape' long reels of yellow paper, like the old bus tickets of yesteryear, with holes stamped through them. Later, we moved over to cards with holes stamped through them and had to carry entire programs around in cardboard boxes; eventually we moved on to the system in use today, The VDU – visual display unit with keyboard. All of this however had to go through the computers located in the basement which was a temperature-controlled environment with the computers running on large tape reels, a far cry from today's laptops etc.

So, the computer application section started with the three of us, Harry, Derek and me. Now Derek and I had not done any programming in FORTRAN (Formula Translation) which Vauxhall had decided to use. All prior work was done in Basic, also our previous programs were very short and technical. No attempt was made to send us on a course, we were expected to 'pick it up as we went along', which is what we did! The programmes were for use by everybody within the organisation and had to be written to be user friendly and structured accordingly.

This blend of mathematical equations for the solution of the reason for the program's existence and the mathematical structuring necessary to sequence everything in the right order, seems to put a lot of people off and several clever draughtsmen said to me that this was the reason that they had not taken up computer programming!

Things improved a lot when Frank Cole; 'Frank the yank' joined us later, he was very helpful to Derek and myself and was already an accomplished programmer. Harry Moss was one of those clever sods, like Pepper, who knew it all! but cannot explain it to others, no good as teachers! Frank introduced us to flow charts, arrays and sub routines etc all of which we had not used in our short technical programs previously.

Frank was a bit of a comic, when asked by Derek how many children did he have, he said three, so Derek said, 'What are they?' Frank replied, 'one of each!' And on another occasion, He had Fred and me standing about during a lunch hour on the playing fields opposite the office, operating one of his home designed machines for demonstrating some phenomena in the earth's magnetic field, as I remember. Only to admit later under questioning that it was all an elaborate hoax! He was a nice man and came to several of our parties at Flitton with his wife Barbara and I kept in touch with him for years after the Vauxhall as we knew it, finished; until his death in the late 90s.

I must say something about the Vauxhall library, it was super, with a competent lady running it. but it was too good to last and started to go downhill when the Americans arrived. The view being taken that a lot of it was old hat and therefore unnecessary, but while it lasted I read as much engine information as possible and when it finally ceased and they started putting papers in the skip, several of us collected as much as we could and I still have lots of MECHE papers, books etc.

The value of a good library cannot be overestimated and it's a disgrace today that the government cuts are threatening public libraries, whilst money is squandered on sporting facilities, leisure centres and other non-educational pastimes, mainly physical skills? Whilst the much more important mind is neglected. You can see this on television clearly, rubbishy programmes with a large percentage of violence, watched by impressionable children in their formulative years – what hope is there for us? The same argument applies, when moneys are freely available for such things as sport, the Olympic Games etc., but withheld from many much more interesting technical projects like the Avro Vulcan, the restoration of an old Victorian pumping station etc, and libraries!

Returning to computer programming, I started to write a programme called 'INDIC', which primarily would work out cylinder pressures and temperatures and from them plots out indicator diagrams for both four stroke Vauxhall engines and two stroke Collett engines! Of course, it was an over ambitious task, hampered by Harry complaining that I should get on with his perceived work! I still have copies of the program. At the same time, I was reading in the *Motorcycle* magazines that a Dr Blair of Queens University, Belfast, Ireland was also writing programs on two-stroke engine development, expansion chamber design etc, and successfully applying this theory to the university's racing motorcycles ridden very successfully by Ray McCulloch – a local Irish rider. so, I gave up on the four-stroke side of the business, concentrating my efforts exclusively to the two-stroke engine.

This was one of the main reasons for my fallout with Harry Moss; he had never really liked Derek and me; we had after all been thrust upon him by Vallas, and by now with the addition of Frank, several other members had arrived all being interviewed by him! What, with personality clashes, no merit pay rises, no promotion and hardly any recognition, by 1979 when Hunting's started to interview for jobs, I took off in that direction, but that's another story!

Around this time, late 1970s, Gordon Blair came to Hitchin to give a lecture on his two stroke developments. Jack, Ted Snook, myself and some others went along. I took the 125cc sleeve valve two stroke engine that Fred and I had just finished but he showed little interest, he would probably be keener if I had taken the exhaust system, after all he was a mathematician rather than an engineer. Don't get me wrong, I realised that my two-stroke programme was always going to be inferior to his for obvious reasons but I thought well, I'll have a go, and I learned a lot with all the researching from the Vauxhall library, talking to people like Phil Pratt; his equations on gas discharge through orifices form a large part of the exhaust and induction phrases!

A bout of tonsillitis laid me low in 1978, I've been suffering from it for many years, but it is getting less intrusive now; and I was forced to take to my bed. Upon reflection I think that the enforced rest, peace and quiet, cut off from the day-to-day activity, enables one to let the imagination roam free. So that all those subconscious memories being stored in the brain can get out. There is too much trivial pressure put on people today such as: changing their energy supplier for a better deal, or searching for the best interest rate, worrying about their pensions etc, all a waste of time!

The argument was reinforced by Desmond Carter in a MECHE paper, of which I have a copy. When he conceived the idea of pressure charging diesel two stroke engines, by using reflected ram from their exhaust systems, i.e. pulse charging. He thought this up and worked out the details whilst travelling on the train from Manchester to London each week, free of office pressures. Mind you, I don't think it would be possible today with mobile phones, overcrowding and noise etc. During this week I invented 'Phased Transfer'. Well, I'm not entirely sure but I have subsequently spent much time checking but have so far not found anything quite like it! I have not patented it, merely sat on it for 30 years, until it was published in *YOWL*, the Scott owner's magazine in 2008, thus rendering it now unpatentable. Most of my friends knew about it but the reason for not taking any action was due to the complexity and cost of a 4-cylinder engine and our (Fred and I) involvement with the sleeve valve engine, all of which are single cylinders.

With only my wage coming in, we were stretched financially during these years with things like mortgages, school fees – later helped by my father, running vehicles and living generally! All vehicle and house repairs were done by me. I never took a car to a garage for repair, doing all of the work in house, in the pit that was dug in the barn! About this time, Steve Linsdell approached me with the idea of sponsoring him on the Scott/Norton, and as he had been racing very successfully on his own Royal Enfields for some months; it seemed like a good idea.

I stripped the bike down and rebuilt it with a Rudge piston of 85mm bore, giving an uprated capacity of 560cc, with lowered compression ratio to run on petrol. A further modification was the fitting of a second spark plug between the pushrod tubes, to be fired by an insulated Magneto, necessary to remain within the vintage regulations. The problem was, push starting, the rotative output of the mag at low revs was insufficient to bridge both plugs, so a starting lever was added that earthed out the Magneto after the first plug had fired! As soon as the engine had picked up, a flick of the lever put it back on to twin plugs. The improvement in performance with two plugs was quite astonishing, as soon as the starting lever was moved, the bike would surge forward and the ignition advance adjusted to suit. The difference between running on one and two plugs was some 15 to 20 degrees BTDC retarded indicating the shorter flame travel.

In fact, I firmly believe in dual ignition for all cylinders over three inches in diameter! As indicated by the mag drop test for all piston-engine aircraft, but of course, primarily it was done for safety reasons. Steve did very well on it, winning his first race at Snetterton, demolishing all the opposition, including the post vintage class up to 1939, going on to further victories throughout the year, until successive mechanical failures, mainly broken crank pins, convinced me to terminate the series. (All of this is in the Scott/Norton notes for 1978.)

There was a man at Vauxhall called George Jones, he used to rent out an old seaman's cottage at Whitstable on the North Kent coast and we had a family holiday there with my father and were visited briefly by John and Rosemary. The upper barn wall on Brook's side collapsed at about this time and Fred and I rebuilt it using beams of Dutch Elm and second-hand bricks left over from the 1976 wall previously described. Penny Horsetail did a painting of the house, which I still have, it is rather on the lines of a caricature.

FTN.8   THE WHITE HORSE, FLITTON   COPYRIGHT FRITH LTD

As we bought it in 1968.

Jack with KTT Velocette, Brook Lane, Flitton - 1968.

Cadwell Park - 1970. L to R John Lane, Miles, me on Scott / Norton.

Brands Hatch - 1971. First on Richard Symes' triumph T100 1939 vintage.

Iris, Sarah, Phyllis, Miles, Flitton - 1972.

Jack's barn in 1972. Miles, Rob, Iris and Sarah.

Crystal Palace - August 1972. Last meeting with
John Lane leading the pack – number 11.

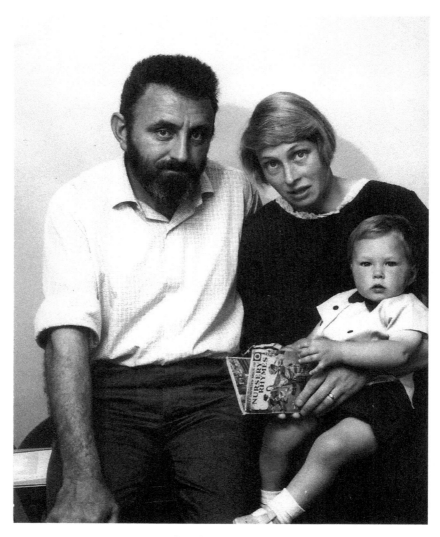

Ray, Jill and Matthew – 1968.

E1 sleeve valve 125cc 2-stroke - 1974. Fred's favourite.

Lane's party in 1976 at Swingate farm. R to L Allison Ascott,
Jill, Robert, unknown, Ray and Anita.

The workers at John Lane's garage in 1976 - a hot summer.

Steve Linsdell with Scott / Norton at Cadwell Park.
Finished first. Sponsored by Rob Collett - owner.

Scott / Norton 490 cc

Built in 1962 by Bob Collett

Winner of the Rex Judd National Vintage Road Race Trophy in 1963, 1964 and 2nd in 1970. It was raced in Vintage events by Bob until the end of 1972.

During this period the fuel used was Straight Methanol. With a Single Central Spark Plug a performance of 100 MPH at 5500 RPM was achieved.

In 1978 it was converted to run on Petrol with Dual Spark Plugs and raced by Steve Linsdell. Many successes were achieved, but the increasing number of Post Vintage events resulted in Mechanical failures.

Out of a total of 90 races it finished on the leaderboard 56% of the time.

Flitton workshop - 1968 to 1992.

179

Chapter Ten

Hunting Engineering – 1979 to 1990

This now brings me to 1979 and the start of the 80s! My decade at Hunting Engineering and the making of new friends. As I said before, my relations with Harry Moss, head of computer applications at Vauxhall, were always 'strained' for want of a better word! When I heard from Tim Lee's wife, Lindy, that Hunting's were interviewing for engineers etc (She ran an employment agency in Ampthill), I took off in that direction, this was the autumn of 1979.

My first interview was at the 'Saracens Head' pub in Towcester, featured in Charles Dickens novel *Pickwick Papers* and was nothing more than an informal chat with the Personnel Department to establish credentials and have a free pint of beer! The second interview at Ampthill was of a technical nature with Frank Bartholomew, the chief engineer and it did not start off well. He, taking the view that engineers in the motor industry were too specialised, i.e. if one got stuck on designing windscreen wipers for instance, then one would always be on windscreen wipers for life, or as long as one stayed with the firm!

However, I pointed out to him that I was writing computer programs for all technical aspects for Vauxhall Motors, and that if he looked at my CV he would see that I had also worked previously in the aircraft industry at De Havilland, AGI, Westland on contract, with Sandall Precision and I had a diploma in Aeronautical Engineering from Chelsea College! That settled it. I was in. I liked Frank, he knew how to use my particular skills, moving me from Department to Department. I must have worked on more projects than almost anyone else in the firm, as you will hear.

My first job was on JP233, the cluster bomb dropped from low flying aircraft as runway busters an antipersonnel mines, used effectively in the Falklands War. I was sent to the Merton centre at Bedford and used to travel up with Rob, Jack's friend from South Africa, who was working at that time attempting to return to England, having I assumed, become disenchanted with the regime there; he bought his family with him, Pearl and son, Nicki; he went to school at Greenfield with Sarah who was about the same age, five or six. They were with Jack for about two years and nearly drove him potty! Especially Pearl who used to run his washing machine nearly every day for the sake of a couple of handkerchiefs!

After a month or so I found the travelling irksome and realised that I was now travelling 10 miles north instead of 10 miles south as previously! So, I complained to Frank B, and he said, "no problem, there are plenty of people here at Ampthill who would rather work in Bedford; give me a couple of weeks to sort it out, we don't want to lose you chaps, as we have gone to some trouble to find you." That is what I call a good manager, not the attitude of many with their 'take it or leave it' dictum.

About this time, Fred and I started on the 500cc single cylinder four stroke sleeve valve engine and it again came from Steve Linsdell in the following manner. He had discovered that a new racing class was being introduced for 360cc two strokes to race against 500cc four strokes! So, I set to and after about three months produced the necessary drawing which I called the E2, engine the second, after the 125cc E1. He found a pattern maker called Nick Moss who lived in Birmingham and was prepared to do the necessary work at reduced cost. So, Fred and I started machining in December 1982 and finished the engine in September 1984, after 752 hours of work! (See photos engine book record for E2).

The whole assembly was to go in Steve's homemade frame, which he designed and built originally for one of his 500cc single cylinder

Royal Enfield engines. it was based on a Bob McIntyre idea for an AJS 7R installation. (See P.E. Irving's book – *Motorcycle Engineering*). this took a further 139 hours and on 1 December 1984 it roared up and down Flitton High Street, illegally!

At the same time, a further article appeared in *Motorcycle Weekly*, extolling the virtues of the sleeve valve engine! And concentrating on the final performance and evaluation of the 125cc E1, with some reference to the aforementioned 500cc racer.

Through all of these years, we used to visit Uncle Tom and Auntie Louie at Bugbrooke with the children, usually on a Sunday afternoon, where they would play in the back garden alongside Tom's large greenhouse where he used, as before, to grow all kinds of different vegetables beautifully and give plenty to take home! But as in all things, it came to an end when first of all Auntie Louie died in 1981, followed by Tom in 1983 – end of an era! Iris and I went to see Uncle Tom 24-hours before he died, aged 95 in Daventry hospital in 1983. We found him in bed and still lucid. He said, "I've had a good, interesting life with no regrets," and as we left after saying goodbye, I looked round and saw him putting on his glasses and picking up *The Times* newspaper! Astonishing! 24-hours later he was dead.

In many ways, Tom had a greater influence on my life than my dear father. For a start he was an engineer (electrical) and had strong hobbies and interests and seemed to talk my language, whereas my father and I had little in common really. He, having absolutely no knowledge of engineering or practical matters at all, my mother did all the household repairs, fuses, switches, electrical items, etc. Tom and I discussed issues of politics, religion, life, etc, and I remember him saying, "the greatest threat the human race faces is overpopulation!" Well, it's doubled at least since then!

He regarded John and me as second sons after Bernard, both of us being engineers whereas Bernard was a disappointment, not being very bright and suffering from mental health problems. Shame!

Meanwhile, the bike was undergoing development in order to prepare it for its first dynamometer test at Leon's workshop. Leon was a friend of Steve's and his works were situated somewhere near Nottingham, if I remember correctly.

The first test revealed 27.6 brake horsepower at 5,000 rpm and eventually after many modifications including the fitting of three MK1, Concentric Amal Carburettor's, twin spark plugs and a sleeve contraction ring to control the excess oil from working itself up the outside of the sleeve and into the exhaust ports, it eventually developed 39 BHP at 7,500 rpm with the specific fuel consumption of .66 points/BHP/hour (see notes) on 26 June 1986.

We did three trips to Leon's Dyno, and then a terrible thing happened. He was killed on his way to work, in his car! Apparently, a large road roller on a trailer, overturned and crushed his car! Awful. However, Fred and I had seen the need to have our own Dyno and the 125cc one was obviously no good to take a complete motorcycle, so with the use of a couple of RSJ's, supplied by Dave Tebbut, forming the base assembly, we fitted the Telmar retarder and the necessary line shafting so that a bike could be clamped at right angles with the rear chain coupled to a drive sprocket (see photos). With this arrangement, we did a further eight runs with many modifications, but no substantial power gains occurred, so Steve decided that 40-odd horsepower would not be enough to be competitive, so that was that!

During all of this time, we tried 1-2-3 carburettor's and 1-2-3 sparking plugs and about six different types of cylinder heads with and without squish, and the appropriate pistons to suit. Many of these ideas were later sent and discussed with Headley Cox in America (see letters to H. Cox etc) he was building and developing his own sleeve valve four stroke engines and had been an early inspiration to me when his first sleeve valve article appeared in the motorcycle in 1951, a 250cc single.

An amusing incident occurred during this time, when as previously mentioned, the bike was tested illegally. Steve disappeared on it and was gone for some time. Fred and I waited and eventually heard it coming from the Wardhedges end of Flitton at high speed. He rode up the High Street and straight into the barn shouting, "get the doors shut quick", so we did. The three of us peered out through the cracks to see a police car go by towards Greenfield and then come back again several times. Later Steve said that he had passed them in a layby on the Ampthill bypass at speed and decided to outrun them! Close shave, we would all have been in trouble!

On one of our Dyno test runs one Saturday afternoon, we had a visit from a woman who said that exhaust smoke was filling her house up and would we stop, of course we did! Apparently, she was upstairs when smoke started coming up the stairs. she naturally thought the house was on fire but upon following the trail of smoke, it led to our, open to the road, front doors! It was very odd because unlike the Brooks next door problem when testing the 125cc two stroke on our first Dyno; this smoke, due to freak wind conditions, must have spiralled down the road and somehow turned through 90 degrees and gone in through her open front door – very odd!

This now brings us to 1989 when work on the bike more or less stopped for the following reasons: -

1. Having to take control of my father's affairs with repeated trips to Wallington each weekend.
2. Marital problems with Iris – leading to eventual divorce.
3. Redundancy from Huntings and working under contract at the Royal Aircraft Establishment at Thurleigh.
4. Moving to a new house from Flitton to Stoney Stanton and subsequently remarriage to Jill. 1992/1993.

However, when possible some work was undertaken during this time 1989 to 1994, although not logged. Work again started and this involved converting it from a potential racing machine into a road bike: a decision taken by Fred and me in order to get some practical use out of the engine – unlike the previous 125 E1. In order to do this, we reverted to a single carburettor, later twin carbs; we retained the twin plug setup, using a double ended coil. We fitted Norton front forks, a Seeley fuel tank, a Royal Enfield back wheel, a Yamaha alternator, all loaned by Steve, including his original frame! Generous indeed! MZ headlamp, saddle and rear shock absorbers and front mudguards from our spare stock and John Urwin welded up a primary belt and rear chain guards. (see photos). this now brings us to the end of the 1980s, so we will return to the early part of the decade to pursue other topics!

In the early part of the decade, just before I started at Huntings, the phase transfer idea and drawings were taken by me and Kevin Wretton to 'Whadden', a firm in Whadden near Beddington in Surrey. This organisation was, at that time, racing 250cc motorcycles and Wretton was, I believe, riding for them. He was an ex-Vauxhall apprentice and had come round earlier when we were testing the 125cc engine. Unfortunately, they lacked the vision to see its potential, if any. They also made the mistake of asking Dr Joe Ehrlich's opinion; he was, I believe their consultant at this time. Well, the reply was obvious! If it wasn't Joe's idea, then it was no good; ask Ted Snook! Joe never invented anything, he just pinched other people's ideas according to Ted who worked for him for many years! However, John Lane pointed out after reading Mick Walker's book on Derek Minter (page 138) that Joe, quote: – 'also sketched out a 500cc liquid cooled Rotary disc induction square four two stroke': unquote!

It would be interesting to know what the transfer system was. Apparently, drawings were made! Ted Snook wouldn't know about it because he had left Joe at this time to come and work at Vauxhall's with Gerry Fitzpatrick and myself. A second attempt was made later, just after I had started at Hunting's in 1981. When I discovered that they had bought up a firm called 'Fieldhouse' at Anstey, Leicester, which specialised in two stroke engines (see their brochure). I wrote and suggested to them that I would do a four-cylinder phased transfer engine based on two of their 380cc twin cylinder engines, back-to-back, called Sting 380W.

Reference to the correspondence will show that I received favourable results and they seemed interested. I had an ulterior motive of course, revealed in later communications of 1983, whereby if all went well, I could transfer from Ampthill to Anstey's engine division still working under the Hunting umbrella. But Alas it all came to nowt as usual. Whether Ampthill management would have let me go anyway, I don't know, and I never found out. What I think happened was that Hunting's closed it all down, retaining only the premises; it's called 'asset stripping' and I've mentioned it before!

To finish with the early attempts to build a phased transfer engine, I did some drawing based on the 250cc Yamaha engine, with the idea that Steve, now being a Yamaha agent would supply the necessary bits. The idea got around and Dave Draper, a fellow racer at Hunting's, told me of a man called Gerry Pell who lived at Hackleton, Northamptonshire, not far from Horton, where John Easton lived, who supplied and repaired Yamaha cranks etc. So, I paid him a visit and he agreed to make and supply two special 90-degree crankshafts for £500 but it was all a bit premature and this time it fell apart due to me being financially strapped; only to revive later in 2008 as a Scott/Norton MK2. E9!

These early drawings are still available in my collection of original prints. The first one uses overlapping Rotary disc valves for induction; whereas the second uses reed valves for the same function and is therefore slightly simpler and was drawn later as reed valve technology developed. They were all based on Yamaha crankshafts with the later one over bored to give 600cc capacity. Also, they are not 'E' numbered like the rest, as they never came to fruition in any motorcycle; nevertheless, it could all come to life again, if say, combined with the E10 vee engine project by someone who considers it worthwhile!

Many things happened in the early 80s and they are listed in chronological order as best I can remember them! A lot of them occurred simultaneously; however here goes: -

Ten years after we moved in, dry rot was discovered in the panelling of the snug bar which I removed and had replastered, also constructing bookshelves, a new fireplace and raised brick hearth with a gravestone from the old, ruined, Clophill church! This enabled us to continue to have bigger log fires in winter as previously described. Chris Plummer died also around this time, quite suddenly of a heart attack. He had just finished renting a room or two to Tim Lee, so he was on his own when it happened. It took everybody by surprise as he seemed such an energetic chap, making and playing his own banjos etc. He left a wife, Valerie, divorced I think, and two girls Mandy and Wendy, the latter being a great friend of Sarah's whom she still keeps in touch with.

Steve and John Irwin got married on the same day in 1984 and we went to Steve's because he had asked us first! It was held at Flitwick church and the reception at the Flitwick cricket ground club house next to his fathers' nursery. Meanwhile, I was visiting Wallington to see my father, usually on a motorcycle at weekends and at the same time catching up with my old pals like Martin Fenney, Ernie Evans, Geoff Gunter, all of whom still lived in the area and on Saturday evening would visit Anita and husband Roy at Redhill to talk about bikes, they both being keen riders of motorcycles.

At this time, a series of books on Croydon Airport were published by the Sutton libraries; I still have them and my father sent me several copies as and when they were published, they were especially interesting to me as, of course, we lived on the edge so to speak and several of the incidents quoted in the past I personally witnessed! Sarah, at this time, started at Clarendon School in Haynes, with the help of father paying some of the fees! This, more or less, coincided with Miles finishing at Bedford School with his O levels in 1983 and starting as an apprentice at Huntings.

It happened thus: – every year at that time some firms offered a certain number of apprenticeships, remember these were the days before the universities convinced everybody that they we're the only way to success! Things have changed a bit now. However, Hunting's were offering six places and I believe over 100 applied. Miles decided to have a go and just before his appointed interview I said to him, "take the steam engine you made at school, it will show them what you are capable of." He was very reluctant, probably embarrassed but I persevered saying, "You've got no real CV for obvious reasons, this will help."

To this day, I am convinced that this was possibly the main reason he got in. He was the only boy who brought an example of work done! Shows initiative! Huntings had an arrangement with Cranfield University, whereby they both had access to each other's libraries, and this proved very beneficial to me as I visited there once a week in the evenings to borrow books on engines just like the old days at Vauxhall! Although I had left Vauxhall, I still kept in touch and belonged to the UFO group, which met once a month usually at the 'Stone Jug' pub in Clophill. We went on several long weekends with them to the New Forest, Peak District and Norfolk, etc, plus walks and visits to Newnam barn dances.

In the autumn, we made cider, apple wine, etc under Derek Keep's surveillance. (He was also a member of the Vauxhall vintner's society) using Jack's apples. In order to accomplish this, Derek Fensome made a wooden 'Cheese', which is a wooden structure in which the shredded apples, wrapped in muslin, are placed and then pressed to squeeze out the juice. for this, Fred and I manhandled the ex-John Collett hydraulic rubber press (which I still have) round to Jack's garage where for nearly a whole day, the combined efforts of the UFO members turned apples into cider etc and eventually to pee! Sorry.

In our case the cider was stored with the homemade beer in our cellar – perfect conditions, in large two-pint screw top bottles, in fact, the original cider bottles! It was rather 'sharp' but blended with some commercial Apple juice – very palatable! However, a visit from John and Rosemary, who after sampling it said her father, he was a farmer, would have enjoyed it exactly as it was! He came from Dorset! Also, at this time Iris had the children christened by our local vicar with a ceremony in the garden afterwards. Not at all to my taste, but I was overruled, as the children had agreed!

Initially, when they were born, they were not christened, and I remember when Miles was born my mother raised her eyebrows and saying, "not christened?"

I think now, 2015, in retrospect that Iris has probably changed her viewpoint again, who knows? One evening, just as Iris and I were about to set off for a social do, the wind got up, there was a loud crash upstairs and it turned out that one of the large chimney pots at the West End of the house had been blown off, crashed onto the roof bringing down a shower of tiles!

I had it repaired on the house insurance, estimated at £700 but had an argument with the builder whom, in my considered opinion, had not done enough work to justify that amount of money. When I contacted the insurance company they weren't interested – typical. That's why our premiums are so high, same with motor vehicles. Miles knows; when he and Pat (first wife) were rammed in Bedford, they wrote his bike off, paid him out, he bought it back for a song, after asking them what they would do with the machine, to be told, "We will scrap it". He repaired it; only minor damage to the rear and is still riding it today! Or was, he sold it to John Martin.

Whilst all of these events were taking place, 'Roly' our only remaining cat, was churning out kittens, in the early years at the rate of about 10 a year, a batch in the spring and another in the autumn! Incredible. She must have given birth to about 100 kittens in her long 18-year life! People used to advise me to have her spayed, but we mostly managed to find homes for them; in fact, towards the latter years when she was having fewer, we had a waiting list, due to the fact that so many female cats had been 'done'. In fact, she had her last litter here at Stanton in Mick's barn and it was given to Jill's friend Jeanette who called him 'Henry'.

Jack's Rudge came in for some restoration work at this time. he decided to have it rebuilt as it had done many thousands of miles and he felt that it was getting a bit tired. The engine was given to Steve and the gearbox sent off to Mervin Stratford, of vintage racing fame. Jack and I worked on the cycle bits, I remember re bushing the front forks and making some new gearbox fixing bolts etc. He bought new silencers but left the original exhaust pipes as the chrome was still good. It now went on to do many more miles and this is the condition it is in now! Not concurs exactly, but very original, so much so that a few years ago, two chaps turned up at his house who were writing a book on Rudge's and we're looking for an original model for each year, as so many of them had been 'tarted up'! Jacks is the last of the line in 1939!

In November 1985, I went to the opticians in Ampthill for an eye test, as I had been having trouble reading and sure enough, they recommended glasses as you would expect! It is an age-related problem apparently as the curvature of the eye lens changes causing near vision to be out of focus. They cost me £18 and I'm still using them 30 years on, in spite of well-meaning person's suggesting frequent eye tests – they do like to spend their money!

Phyllis's, husband Dan, died at this time in somewhat tragic circumstances, leaving her with two 8- and 10-year-old lads, Peter and Paul. Apparently, she came down in the morning to find him dead on the sofa, he had taken an overdose! Difficult to imagine how awful it must have been for the poor girl. I don't blame Dan, he was the victim of circumstances – an unwanted child, poorly educated, a succession of dead-end jobs and eventual unemployment. Tragic! My father who had helped financially for many years was not told and did not attend the funeral, as Iris and

I did. We went in our new Austin Maxi car which we had bought for £100 the first time that I had paid more than double figures for any car! The long-suffering Vauxhall Dormobile that Jack and I had shared went to John Lane who was also running a similar model for his self-employment job of taking the local school children to school.

This was his last real job (Lyn was still working in the NHS) and he went on to pioneer house husbandry where as you know the rules are reversed, the wife becomes the wage earner and the husband: the wife! Quite common now but in those days few and far between, I went on to follow his example later when I married Jill!

In 1986, I went to the Isle of Man TT as Steve's mechanic where he was racing a 600cc four stroke four-cylinder Yamaha which was a new model they were introducing at the time, and a 600cc four stroke single cylinder four valve machine, also newly introduced which he was not racing but showing it off to the public. He was highly regarded by the Yamaha dealership quite rightly so, for his racing exploits on their behalf.

We started off at 5:00 am in the morning I offered to drive his van as I could see that he was very tired after being up all night working on the bike! He refused, all went well until we got on the M1 heading to Liverpool when I notice that we were drifting slowly into the slow lane and onto the hard shoulder! Not much traffic about at this time in the morning, fortunately. I reached out and gently push the steering wheel straight again; after it happened a second time, he saw sense and retired into the back for a kip.

We had a rough crossing and had to wait until 9:00 pm for our evening meal as the next boat had to stand off from Douglas Harbour for several hours before the guests could arrive. He put in respectable practise times, but it became apparent that the machine was not going to finish the race without a pit stop for fuel!

As his bike was slower than some of the fancied runners, he decided the only way to win was to go non-stop. So we spent a lot of time getting the last ounce of fuel into the tank by removing all air space and leaning down the mixture etc as being a production machine – no modifications were allowed! In the race he finished second and could have won if he had kept going as before on laps one and two but he got worried that he was going to run out of fuel, so he backed off slightly on the last lap only to be pipped at the post! One of his many 'almost' wins in the Isle of Man races! On several occasions, I took the single cylinder 600cc Yamaha around the island and Steve said, 'would I like to ride it home' as he was staying on for a second week, I think? I had ridden both bikes and although the four cylinder had obviously more power, I preferred the more docile single.

Just after this, Miles decided to upgrade his road motorcycle (he had had a series of small machines like a 125cc MZ, 250cc Kawasaki etc and he decided he liked the new 600cc four-cylinder Yamaha, so he approached Steve, who offered him at a reduced price the bike that he had ridden in the TT but Miles refused. I understand that, he wanted as his first real bike – a new one! And so, it came to pass, on the hire purchase of course but he was now earning a wage at Huntings.

One of his first rides was to come down to one of the Lanes parties at Walderslade where the bike was duly admired and he and I returned home via my father's house in Wallington, having crossed London in heavy rainy conditions, I was on a 250cc MZ, he did well not to drop it on some of the slippery roads with so little experience with such a large machine!

Visits to my father were becoming more frequent now and I brought him up to Flitton at Christmas, Easter and for summer Holidays. On one occasion on 16 October 1987 the great hurricane struck southern England, doing extensive damage (see *In the Wake of the Hurricane* by Bob Ogley). I went down the following day amid a scene of devastation having to make several detours to avoid blocked roads! Trees were uprooted and houses damaged in my fathers' road, but he had slept right through it. It arrived in the early hours of the morning, fortunately not a lot of people were about but it caused commuter mayhem; the Lanes were without power for several days and their phone lines were down. See book entitled *In the Wake of the Hurricane* by Bob Ogley, of which I have a copy.

During one of these visits, I removed the remaining family silver from the national Westminster bank vaults and left it with the Fenny's for a week whilst I went back for the car the following weekend. My intentions were to get everything of value out of the house in case of burglary or return of the crooked antique dealers from Brighton who had already swindled him out of a Grandmother Clock that stood at the top of the stairs, the hall umbrella stand, nested tables and a glass fronted display cabinet that used to contain some of the family silver. He must have had enough sense to put it in the bank vault! He also had visits from cowboy builders who would talk him into various jobs, do them badly and charge the earth – crooks preying on the vulnerable elderly! All this stopped when the Bakers took over his care in 1989, more of this later.

In the August of 1987, we, that is Dad, Iris, Sarah, and I went on holiday to Dinas Mawaddwy for a week in a hotel. We toured the area and made a start on Cader Idris but did not get far; it was during this week that the Hungerford's gun massacre took place, later followed by Dunblane, which resulted in the banning of handguns in this country! (All photographs of this holiday and all others up to 2019 can be seen in our collection).

My cousin Bernard who was married to Bessie, died in July 1988 and Iris and I attended his funeral in Northampton. The following year Aunty Alice, John's mother also died, leaving Phyllis and me a substantial holding in Shell shares, which was very surprising – I didn't think I was one of her favourites!

Linda Evans now decided all of us OWs (Old Whitgiftians) Should attend a school reunion being organised for boys who were there between 1947 and 1952. (See reunion dinner list and photograph). Richard Symes, R. Evans, G. Gunter, Barry Roach and myself are shown but two others from our form are on the list, Clive Ballard who was previously mentioned in the escapades of Ballard's Wharfe, And Derek Tullett who later went on to found an international financial brokerage firm that is now part of 'Tullett Prebon' The substantial city firm (see *Who Runs Britain?* by Robert Peston!) He is now a millionaire of course; all by just moving other peoples' money around. Jeff Gunter was best man at his wedding and was later told not to visit him again – now that he was successful – cheeky sod.

Meanwhile back at Flitton Engineering, 'Flittenge' as it was known, plans were afoot to extend the workshop facilities. A pit was dug, and an overhead gantry installed which was capable of half a tonne lifting capacity, serving both the pit and the newly installed 12-inch Churchill lathe which had arrived with a Jones and Shipman tool post grinder, 1906 vintage, one of the first they made and a hydraulic rubber press. All of these were donated by John Collett who was closing down the injection firm at South Norwood prior to becoming a gentleman farmer at South Nutfield.

In order to pick them up and bring them back we borrowed Steve's van, which was by no means new and on the way back with all this overload, we had some hairy moments, like going down Norwood Hill in low gear, seeing pedestrians crossing, praying that no one would step out to certain injury or death! When we finally got home and unloaded everything it was found that the rear floor had become detached from the mudguard supporting structure! The cylindrical grinder at two tonnes and the fly press plus other tools etc, were too heavy for a return trip, so Steve hired a vehicle and these items were installed in his workshop, which at that time was in the rear of Flitwick motorcycles.

In the August of 1988, Dad and I visited Baildon for what was for him, a final visit. We stayed at a hotel in Ilkley on the moors edge and also toured the local area, visiting Harrogate, Brimham Rocks and Saltaire, among others (see photos) and I remember him saying at the time, "Well, I never thought I'd come back here!" It was on this visit that I went to David Berry's old house and got a forwarding address as I have previously mentioned. We went and looked at 30, Ferncliffe Drive and I walked the old route down the 'bank' to Sandall Road School and had a look around the Glen cable railway of 1890s, very nostalgic! It was exactly 30 years later, 2019, that I was to make the same trip with Miles, Sarah, Peter, Paul and Jenny but I didn't say the same thing as I hope to return again soon!

We come now to a period in my life that can only be described as distressing, i.e. separation! Things had not been going well with the marriage, I will not elaborate too much on this but one Sunday morning I came downstairs to find a note on the door mat informing me that she had left to go and live in Luton with a man called Russell! They had met some months previously at the Cross Keys pub in Pulloxhill, where we used to go on a Sunday night to listen to traditional jazz. It lumbered on like this for the next four years with Sarah living there during the week, as she was now attending sixth form college to get her A levels in Luton and Miles varying his time between living with me and his girlfriend Sarah up at Wardhedges!

Also, during this. My father was becoming less able and I received several calls from the social services in Wallington to render help; so, I began the period, previously mentioned whereby I had to go down to see him, more or less every weekend. It was on one of these visits on my MZ Motorcycle that I noticed his car was not in the garage and asked him where it was to be told, 'it's in the garage', so we went and looked! I asked him when he last used it and he said, 'a couple of days ago to go shopping.' the resultant trip around Wallington located it in the library car park, so I returned and asked Mr Baker who lived opposite to drive me back and I followed in dad's car – crisis over, but it shows how forgetful he was getting.

Eventually, I decided to bring him to Flitton and try to get him in somewhere like the Cedars, where Mrs Crawley had been taken, when Jack found he could no longer cope with her care. Before we left, however the Bakers came over and said they were prepared to look after him. Mrs Baker would give up her job in Croydon and I would reimburse her out of father's income. I should have taken her up there and then but thought my original idea was a better one!

However, I struggled on for some weeks going to work in the morning as usual, leaving Dad in bed, then returning at lunchtime to get him up, give him his breakfast and park him in front of the television until I returned for the evening meal. It took more than the lunch hour to achieve this and I was always late back in the afternoon! Meanwhile, I visited the Ampthill social services and explained the situation. I foolishly thought they would respond quickly, what a joke, it shows how naive I was! Eventually, common sense prevailed, and I took him back to Wallington and the Bakers who did an excellent job of looking after him until he became doubly incontinent. he was then transferred to a care home in Sutton for about three months until he died on 6 April 1993, just beating the tax man! As you would expect!

During this time, he signed the enduring power of attorney papers, giving me control of all his affairs. Iris's father also died around this time, also in Sutton. He had been ailing for some time, living on his own as his wife had gone some years previously, suffering from Alzheimer's. Not so for the old man, he was sharp right up until the end.

Once again, Steve asked me to mechanic for him. This time in Northern Ireland for the Northwest 200, held on the triangular circuit near Portstewart. Jack also came along as 'backup' mechanic and the three of us sailed to Belfast and then drove up the very beautiful Antrim coast road, well worth a visit. We weren't a lot of use as mechanics, but we rushed around the pits helping with the organising. I was wearing my American army

officer's raincoat and my old army beret, a real target for the
IRA and this was brought home to me. Eventually, when a small
boy sprung to attention as I passed gave a Nazi salute saying,
'Heil Hitler!'

After the racing, in which I think, Steve finished about six against
a strong international field. We attended a post-race prize giving
and general celebrations, dancing, drinking etc, this is where our
troubles began. Halfway through the evening Steve was taken
away by a group of well-meaning Irishman, all smoking cigars, for
some rider's celebration or other! There was no sign of him gone
midnight when Jack and I decided to return to our hotel; so, we
went to bed. We were woken in what seemed the early hours of
the morning by the sound of violent retching from the bathroom!

The long and short of it was, that we had to get him and the van
and ourselves to Belfast for the morning sailing to Larne in
Scotland and hence home again. We dumped him in the back with
the bikes, I drove with several stops for him to get out and vomit
at the roadside, supported by Jack. We finally reached the boat
where he rushed off to the loo! Whilst Jack and I washed down
the side of the van, where he had spewed over it. He spent most of
the voyage in the toilets and on one occasion when we were in the
canteen having lunch, we saw him leaning over the side of the
vessel. When on the boat, Jack and I pretended we didn't know
him. Getting him back in the van as unobtrusively as possible. He
finally came round somewhere near Northampton on the M1 to
ask, where we were and saying, "Never let me go off with drunken
Irishman again!"

We now come to the first of our long period of holidays starting
with Raera in August 1989 and finishing in Norfolk, October
2013. (See the comprehensive list of dates and personal, plus
photographs). It was, as usual, organised by good old Lyn, after
she, John, Jill, Allen, Jacqueline and Paul had been the previous
year; their first week spent in Aviemore. We went in three groups
in convoy; Lyn and John driving, Tom and Sylvia in the Lanes' car,

Jill driving, Alan, Paul, Jacqueline and Sarah in Jill's Renault, with Jack and me in a hired van carrying two MZ's and Johns CSI Norton!

We stopped briefly for lunch on Ilkley Moor, arriving late afternoon at Raera. We were there for two weeks approximately, using the bikes to run around the area and on the second week to visit Norman and Edna in Scourie, where they were on holiday in their caravan. Upon arrival in Scourie, we secured bed and breakfast in an old folks' farmhouse, but they only had one room with two single beds. So, I spent the night on the floor in between John and Jack who both snored. Soon as one had stopped the other started. Dreadful!

We went on up to Durness, looked at the Smoo caves etc, lost Jack for a while then returned via Lairg to Ullapool where we had another B&B before returning to Raera. All the others Meanwhile, had gone off to tour Mull, Iona, Staffa, Fingall's Cave and Barra's coastline. It was during this time taking Jill out on the pillion of the MZ that's our romance began to flourish, finally cemented the following October when Lyn and John bought me up again to Stanton for the harvest festival concert starring Tom Goodluck; a very funny man, good teacher and brilliant at repartee.

Meanwhile, at work as previously mentioned, I was involved in many projects, initially after my return to Ampthill; VJ291. Which was similar to JP233 in that it was dropped from a low flying aircraft, flew for a while, before the tailfins rotated through 45 degrees to give it spin. To enable the munitions to centrifuge out through detachable side panels. My contribution here was to come up with suitable mechanisms to erect and rotate these fins, of which there were four. Later, it was taken over by the newly opened Bristol office which also became involved with my next project – an underwater submarine type mine launcher.

I was told it had to be launched at constant velocity to minimise the G-forces. They had already had a launcher which used elastic bungees, but it now needed updating and if possible, could we still use the elastic Propulsion as they had a large store of these. After some head scratching, I came up with a 'fuse' type mechanism, similar to that used in clocks, whereby a cable attached to the submarine was rotated by the 'fuse' shaft driven by the extended bungees. The 'fuse' wheels of which there were two, were made on one of the newly programable CNC milling machines, recently installed – commonplace now but the forefront of technology then. Involving the employment of suitably qualified personal, which I believe took the total number of staff to over 2,500, the highest it ever achieved, I believe! Miles will tell you.

This was at the start of the Thatcher Tory Government, before stringent budget cuts came into force, when money seemed to be no object and the taxpayer funded everything, especially in the defence industry! Eventually however, the job was done by the Bristol offices pneumatic launcher which they had been designing unbeknown to me! It certainly costs more than my 'belt and braces' approach, and I felt a bit 'peeved', because I'd been told to use elastic bands and keep costs to a minimum!

Next up was MLRS (Multi Launch Rocket System) and was an international project with Honeywell of America. Dornier of Germany, TRT of France and Hunting's for the UK. Again, money seemed to be no object, many jollies' i.e. funded trips, generally to meetings, held away from the firm so that lucrative travel and hotel expenses could be claimed! It goes on to a greater or lesser degree in all firms and organisations, unfortunately! I went to Paris, travelled on the Metro, boat trip on the Seine, and later to Friedrichagen Germany where the great Zeppelin's were originally made, all very interesting. Finally, however, they wanted me to go to the USA, somewhere near Chicago as I remember, where the temperature was minus 40 degrees this winter! Ridiculous, not for me!

My involvement on the project was the ejection of submunitions from the main rocket booster and to this end, I had invented the reversing diaphragm – a thin metal sheet that folded behind the munitions, and was blown outwards by either compressed air or hot gas, I have a patent on this. So, I said to Newman, the project manager, there was no point in me going to the States, interrupting the test work on this design at Thurleigh and in the Bovril, just tell me the meeting results when you get back! Thing was, that most of the meetings consisted of administration chatter and feasibility studies, little relevance to a practical engineer like myself! I think also that the Americans shared my views, they like to see hardware.

Later on, during one of the Hunting reorganisations, a section on ejection systems was set up under Ken Beaumont with Bob Ward and myself, instigated by the success of the reversing diaphragm; later joined by John Evans of Counter mass gun fame? Ask John Urwin! During my time with this group, I went on to invent the 'letterbox and flowerpot bags' funded by a new contract with the RAE Farnborough. They operate by extension from the folded state through a slot either straight, like a letterbox or round in the case of the flowerpot type. Initiated by either hot gas or compressed air as the reversing diaphragm and the BL755 bags, which were already in use by the services.

I have mentioned Bob Ward, I came across him on the Firecracker project – you don't remember the firecracker? Well, I will refresh your memory. The RAF at this time, decided to replace the aging Hunting Provost training aircraft with something new, just as had previously been done in the 40s with the Tiger moth; that shows how old I am! So as usual, the government instead of keeping it inhouse, and supporting the home industries put out a tender with the whole world, not just the commonwealth or the EU, oh no! As a result, several countries, France, Germany and Brazil I remember, there may have been others who all went for it.

Well, ours was called 'Firecracker' and it was a two-seater, turbo propelled, low wing monoplane and a special office was set up in great secrecy with a team of about twelve of us, run by Bill Blakey, the deputy chief engineer under Frank Bartholomew. I wasn't on the project for long, it lasted for about a year before it crashed! A political decision of course, finished it like all the others I've previously mentioned; TSR2, blue streak, Rotodyne, even Concorde, if the French had let us! Reference the book *Empire of the Clouds* by James Hamilton-Peterson, when Britain's aircraft ruled the world.

It was given to the Brazilians and called Tucano – It's still in service now, as far as I know, And the political aspect? Some deal involving coffee beans as I remember; all in the newspapers of the day. After this came a mine launcher project in which three of us designers were involved in competition; the aim was to launch as many mines as possible in the shortest time, a distance of about half a mile, the mines were small diameter about six inches and four inches long.

There were such launchers around, powered by explosive cartridges but explosives are expensive when used in these quantities, one per mine. Geoff Williams, an older and experienced designer continued with explosives, trying to find ways of reducing costs – by launching mines in groups etc. Another chap, who's name I forget, was going to use compressed air and I decided on a liquid fuel (petrol) launcher.

It involved a hopper to feed the mines into the breach, where upon a spark plug would ignite the fuel/air mixture to hurl the mine up a six-foot barrel, to the distance required. The fuel was sprayed into the barrel by a small electric driven pump between successive firings. I calculated that it would fire 360 mines per gallon at whatever rate the hopper could deliver, the hoppers were not our responsibility, the price of petrol at that time was around £1 per gallon! So, 'mine' was without doubt the cheapest option!

The idea was not so crazy as it may seem, because liquid propellant guns were around at that time. The compressed air gun was second cheapest I remember but the heavy pump and drive unit, powered by diesel engine, to get to the required pressure for mine launch, delayed departure between launches. Thus, reducing the rate and the whole contraption was expensive to make. However as usual, the Ministry unwilling to take any risks, decided to rely on the tried and tested cartridge solution, so it all came to nought! Very interesting though, I enjoyed it.

Whilst on the subject of mines, I was moved up to building seven where the LAW project was just coming to an end on another mine laying project; this time to put a much larger mine underground! The army had such a machine that operated by drilling a hole in the ground, dropping the mine in, then refilling the hole! As you can imagine a time-consuming operation, not what you want with the enemy approaching!

Once again, some lateral thinking was required and I remember seeing a machine years ago, laying turves of grass, dragged along by a tractor so I paid a visit to an organisation in Westoning I think, who had such a machine and examined it in detail.

The operation was somewhat easier than I required as the machine acted like a large woodworker's plane, in that it skimmed of the top layer of grass/soil with a horizontal cutting blade and dropped turfs via a hopper onto the now exposed earth. My machine would have to not only remove to a much greater depth the topsoil but place a mine and then return the apparently untouched top layer back again!

To this end, I set to and designed a similar arrangement with a forward cutting blade which took the earth up a slope and over a bridge, down a similar slope, thus returning the continuous strip back from whence it came. Whilst the strip was in the air, so to speak, the hopper delivered the mine underneath the bridge and onto the exposed earth, to be recovered as the machine moved forward, and rolled flat by rearward roller.

With my customary enthusiasm, I visited Fort Halstead army barracks, who we were working with and they were very keen on the idea, offering any help we desired. I also spoke to a local farmer who had fields at the back of Hunting's playing fields and he offered the use of his tractor to pull the prototype machine; could I do more? All to no avail Hunting management customary 'enthusiasm' vetoed the manufacture of it, in spite of the fact that it would cost 'peanuts' and could be fitted onto one of our already made, mine disposal trolleys that John Urwin/Bob Minton's section were involved in! I worked for a while in the special systems department on Trident decoys and devised about a dozen different ways of folding/unfolding conical shapes which were well received but the work was purely theoretical and not in my line really!

More interestingly, I got involved in wind tunnel work for RAE Farnborough again. This time, they wanted an adjustable parawing, that they could 'set' to any desired angle for pitch, roll and yaw, without having to either dismantle and reassemble it, or have several wings that already pre-disposed setups! What I came up with was a series of support arms which had ball-joints and adjustable turnbuckles in them, attached to various points on the parasail, so that almost any setup could be achieved quickly by undoing the appropriate items and relocking. This also took me onto Hunting's own wind tunnel which I did some work on. It was only a very small unit, suitable only for submunition flight with objects with no bigger than 4" in diameter, which suited the BL755, VJ291 and JP233 submunitions.

Finally, I have a patent relating to the SARRUT mechanism for placing the submunitions, etc into the airstream in the correct flight altitude, from the flight weapon, so that they could start off correctly by avoiding tumbling etc. The SARRUT mechanism is one of the 'four bar chain' arrangements found in physics books and dates from the nineteenth century; invented by SARRUT who was a Frenchman! By suitably adjusting the length and location of the arms, any attitude within the range of the machine can be achieved. One of the Hunting's mathematical 'boffins', a doctor somebody or other got interested with a little encouragement from me and 'disappeared' for several weeks to write a computer program covering all possibilities – I wonder where it is now!

Whilst all of this was going on, Sarah was playing hockey for the Hunting's team under the guidance of Ross Lewis who was team captain, along with such people as Kathy and Richard Hollis who was also a Whitgiftian, but long after my time. It was sometime around the setting up of the ejection system department and the arrival of John Evans that I was invited to attend an interview in London for a job with an undisclosed firm to be set up; they were headhunting for suitable staff, I don't know who put my name forward or how they got hold of it but several other systems personnel were invited I think. It was a very attractive offer but would have involved travelling etc, so after some thought I declined – better the devil you know!

However surprisingly, it did me a big favour, I was called into the chief engineers office, not Frank Bartholomew now, he had retired to be replaced by Brian Hibbard and he informed me that they had decided to raise my status to senior staff plus a big rise in salary! Just goes to show! From here, I was seconded to Irvin Ltd, Letchworth where I worked on several projects including Cassini the Titan project to send a probe to Saturn and soft land on its largest moon, Titan. Irwin were responsible for the parachutes on the lander named Hugenes, they had been bought up by Hunting's and were specialist in parachutes etc. I still have all the relevant papers on this fascinating project, which has subsequently proved so successful operating in an environment at 180° Celsius!

I wasn't there very long before returning to Ampthill to work again in special systems, having been requested specifically by Dr Cox. So, it came as a complete surprise when I was again called to B. Hibbard's office to be informed that I was to be made redundant! Incredible, this was just before Christmas – December 1990. I'm still not sure why this happened, the general opinion is that I fell foul of one of the directors, who at that time was working at Irvine's and I wasn't respectful enough to him, plus the fact that he lived in Maulden and I used to cut him up on my motorcycle in Ampthill on my way to work in the morning! This I find hard to believe; anyone could be so childish! There was another suggestion and that was that Graham Birchmore and myself, he was also made redundant from the Design Department, were the two highest paid designers at that time and the accountants just simply chose us on that account! Who knows?

On a more cheery note, Jack and I went on the vintage motorcycle Irish rally in the spring of 1990 (see photos) we had talked about doing it for several years and to make it more authentic, we went in traditional wear, i.e. dispatch rider overcoats, army boots in my case, and pudding basin crash hats with MK8 goggles! Both riding girder fork machines, the Velo and Rudge respectively! We went down to Swansea and sailed to Cork with some of the Bedfordshire vintage clubs members – George Wrench, John Biggs, Rod Atkins and surprisingly Len Ore, he used to be a Trojan apprentice back in the 1950s; he later went on to become President of the Vintage Motorcycle Club.

The rally started officially from the Metropol Hotel in Cork for the first three days outings: Finally moving further west for another three days. We used to start out at about 10:00 o'clock in the morning, ride for two hours, stop for lunch, start at 2:00pm, ride another two hours, then dinner at 6:00pm. We would do about 100 miles per day which, for the keen ones was timed, but Jack and I ignored all that, we only went to enjoy ourselves and see the beautiful countryside- –we got a commemoration plate each for competing!

We met some very interesting people as the event is international, I remember on one occasion in Kenmare when some Americans were having trouble with one of their machines and needed to drill some holes, so they broke into this bungalow and used its power supply; when the woman came home they apologised profusely and showered her with money; all ended well. I also remembered the trouble-free rides we had when we returned in the late afternoon, we simply changed into our evening wear and went to tea, whereas most of the other competitors seemed to grovel about with their bikes, goodness knows why. I would have thought that one would have been able to put a machine together to last a week or so without extensive maintenance!

During this time, I phoned Jill periodically to express my love and report on progress, as from Raera, in 1989 we had more or less become unofficially engaged, subject to Jill's final confirmation later. We returned the way we came on a day crossing of about 10-hours, to arrive in Swansea to a pre booked bed and breakfast at about 6:00pm, seeing Dolphins on the way. The Velo (Dotty) was losing power on the return trip and we stripped it down, replacing the main bearings, re-bore, etc as I remember. Jack said it was the first real overhaul the engine had had since he owned it!

Whilst we were over in Ireland, John Lane was doing a Land's End to John O'Groats run with a group of Norton owners, he of course was on the oldest machine and this most economical both on fuel and oil! Turning about 85 miles per gallon and using virtually no oil, so much for progress! Eventually however, his bike conked out near Wick in Caithness, he stripped the engine down in a farmer's barn to find that the timing side main bearing had broken. Being Imperial and of non-recommended size, some delay was experienced before reassembly and the final ride home again!!

To show again what an enterprising chap he is, he started work on a solar panel installation for obtaining a constant supply of 'free' hot water. it must have been one of the first, as no solar panels were commercially available at this time, 1990. He rigged it up on the front of his garage which faces southwest, by soldiering sheets of 18 SWG copper, six foot by three foot together and running the cold water, forced by a 240 Volt pump between them, to a pipe system linked into his bathroom. There was also a backup system involving a 12 Volt water pump plus car battery, which would switch in, in the event of say a power cut, to prevent over pressurising and hence leakage. This water pump is now on the E6, as the whole system was dismantled in 2017, after working successfully for 20 years. He was a pioneer! All homemade, more or less!

By now the Bakers were looking after Dad in his own home, I don't think he ever really understood who they were or what was happening, but he always knew me, unlike John Lane's father who kept on calling him 'Stan'; thought to be an old school friend of his, but he always knew Mike, John's brother, strange! I was working at the Royal Aircraft Establishment at Thurleigh on the design of aircraft installation equipment and an emergency occurred when the Bakers announced that they were going on holiday. Well, I had to rush down to Wallington, bring Dad up to Flitton and take him to Phyllis in Wellingborough, where he stayed with them for a week or so.

Whitgift reunion in 1987. L to R Rob Collett, Barry Roach, Geoff Gunter, Richard Symes, Rod (Ernie) Evans.

My father aged 88 at Moor End, Upper station, Baildon - August 1988.

Raera Scotland - 1989. L to R Jill, Jack, Paul, Tom,
Alan, Jacqueline, Sarah, Lyn, Silvia, John.

Vintage Irish rally in Bedfordshire. V.M.C.C. Team L to R George Wrench,
Jack Crawley, Rob Collett, John Biggs, Rod Atkins. Kenmar hotel - 1990.

Sarah, John Lane and Mrs Brown. The 3 tors 1989 - very wet!

Jack, MZ, John, Norton, Sarah at the Atlantic bridge, Seil Island.

Jack in Kenmare, Ireland - 1990.

Derek and Ray Rowcliffe on the 3 tors circa 1990.

At Glengarrie, Irish rally - 1990. A welcome pint.

Chapter Eleven

Royal Aircraft Establishment – 1991 to 1992

I had started at the RAE in January 1991, through an agency called Howard Delta of Stevenage. I don't know how they found me, I imagined through the headhunting process and I became a contractor working there until June 1992 when the contract ended. The work there was interesting, though not as innovative as Hunting's and I met several interesting people including Ron Hance, a contract draughtsman, whom I still keep in contact with at Christmas. As I was now heavily involved with Jill, going up to Stanton after working with Fred and the engines on Saturday afternoon; staying till early Monday morning; returning to work down the A6 usually on the MZ or Dad's car – Ford Fiesta later, when I took as much light equipment in the boot to clear out the workshop before the old White Horse was sold.

There was a shuttle service between RAE's Thurleigh and Farnborough sites on a daily basis for the interchange of personnel using a twin-engine monoplane, holding about 10 passengers. On two of my flights there were minor incidents, once when the heating failed and it got very chilly and on another occasion upon leaving Farnborough in lashing rain, zero visibility, with a novice pilot, the plane being badly loaded so it was tail heavy! We had a very heavy landing; I thought the undercarriage was coming off!

I had not been there very long when two gentlemen from Hunting's arrived and wanted me to sign a release form, relinquishing my rights to some of my patents, cheeky buggers. It wasn't their fault; they were as embarrassed as me! Well, I refused to sign of course, obviously in their haste to get rid of me they hadn't thought about all my patents! As I was a union member still and they provided a legal service, I contacted them for assistance in the matter and a solicitor took up the case (I still have the correspondence) which went on for several months. The final analysis was that in order to sue them or get any money out of Hunting's, we had to prove that they were themselves, making money from the use of the patents or whosoever they had subcontracted them to! In order to do this, we had to have an 'insider', or I might buy some Hunting's shares and so get an annual report or relevant paperwork.

The solicitor then pointed out that if this were accomplished, I would have to pay patent protection fees for as many years (up to a maximum of 15) as was deemed necessary to protect my patents from exploitation, i.e. prevent anybody else using it. This of course, is progressive as the years rolled by; as by this time, you are supposed to be making money out of it yourself, through manufacture or lease etc, all this, needless to say, would not be covered free by the union! So, I thanked them for their help and that was that. I've heard nothing since, except that John Evans took up the development of the Bags, extending their stroke up to about one metre, against my half metre when I left the firm. This was by curtesy of Bob Ward, my insider!

In the middle of 1991 (June), Jill and I decided to join Jack in Ireland, this would be Jill's first visit there! Jack had already gone ahead on his BMW motorcycle and was staying at his nephew Brian's new bungalow near Kilorglin Co. Kerry. We went in Jill's Renault petrol driven car and toured around the area with Jack, visiting Brian's friend from Vauxhall's who had also immigrated to Ireland, having taken early redundancy/retirement and lived by the riverside in Killorglin, spending his days fishing! Then on to the Dingle Peninsula, where as we were driving along, Jack shouted, "Stop the car," and as we did he leapt out and disappeared into this scrap yard! And then ended up painting the rubbish! He said he was good at that!

We saw the beehive shaped 'houses' similar to the ones I had previously seen on Achill Islands back in the 1960s and of course the Blasket Isles. We also visited the highest pub in Ireland that Jack and I went to visit the previous year on the Irish rally, where the landlords five-year-old daughter, danced on the billiard table (see photos).

A further trip took us to Carrantuohill, the highest mountain in Ireland. I had previously climbed in 1960, as I have previously mentioned. It was on this visit with Jill that I discovered the loss of the record book on the summit; later Jack did a painting of Jill and I starting out to climb it; it is now on our lounge wall! After the first week, we left Jack at Brian's house and went first to the Glen of Aherlow in Tipperary, where we climbed Galtymore the 3,000-foot summit of the Galtee mountains, but the weather was bad, and we saw nothing after about the first 1,000 feet! We had a dangerous descent from the summit, as visibility was near zero and we had come past a large crater, with an overhang and full of water. So, to avoid going over the edge on possibly loose scree, we decided to go back the way we had come. It proved extremely difficult to find the original path, but we managed it eventually although it took some time and care, we thought we would know better next time, but read on! From here we visited the Wicklow mountains on the east coast where the weather improved but did not climb Lugnaquilla another 3,000-foot peak, so we've only done two of Ireland's four 'Munros'. The remaining one mentioned is Mount Brandon on the Dingle Peninsula in 1962.

Meanwhile, preparations were taking place for the impending move to Stanton and marriage, but before this, workshop facilities were needed. Fortunately, Jill had several pig sties at the bottom of her garden which were partly shared by the Copsens, her tenants who lived next door in number 45. So, the local builders, the Groococks, were set on to knock down the intervening walls between each sty, extend the building forwards and upwards, leaving the end sty as a woodshed for the Copsons who only had open fires. At the same time, the washhouse was also extended and renovated to give a much larger storage area. When all this was complete, the main bulk of the Flitton workshop was transported up to Stanton. Steve had a mate, who had a low loader and crane assembly and was commissioned to do the job. No problem was experienced at the Flitton end, but Stanton was another problem entirely; access being severely limited. He parked outside number 51 and lifted the machines over the fence (now a wall) into number 51s front garden Causing much chaos on the busy High Street!

They stood here for a few days until the Groococks were available to drag them across the gardens on scaffold pole rollers and into the new workshop. On the way, the heaviest machine – the Denby Mill at two tonnes fell over on its side on the lawn of number 47 and we struggled to get it back onto the rollers which ran on scaffold planks, Paul came out and gave us a hand, whereas Jill had gone to work to get out of the way! In fact, at this time, Jill was in full-time employment with the NHS; Alan was at Exeter University studying Theology, Jacqueline was at Newcastle University studying medicine and other things!

I left the RAE, stayed on a while in Flitton to complete the sale of the property to Mary Anne Austin and her husband and clear out any remaining items; some furniture was brought up to Stanton by a hired van. I then completed my move to Stanton by living in number 51, until our marriage in March 1993. (ref video of the same!)

We all want back to Raera again with the exception of Jack, Alan, Jacqueline and Sarah, but with Nick, Heather and their two children. From here we visited Mallaig and Fort William on the steam railway, beautiful journey through stunning scenery. We hired a boat to the North End of Jura and walked down to Ardlussa to be picked up again after a barbecue! We also visited the Mull of Kintyre, Easdale Island and did a section of the West Highland Way; Nick, Heather and the children on mountain bikes.

Just after this my father was moved into a home in Sutton, he was now 93 and the Bakers could no longer look after him as he was doubly incontinent! I sold his house (having power of attorney with Phyllis) for about £100,000, below par, as there was a recession. Nigel Lawson was Chancellor of the Exchequer, a disaster! so we lost out on both house sales; house prices have never dropped so low since, there were many repossessions, people unable to pay their mortgages. It has not happened this time, 2008 onwards, as the taxpayer and savers have picked up the tab! In fact, mortgage interests have never been so low 1.5%. In my day, i.e. 1970s, I was paying 16.5% you've never had it so good! Negative equity, what's that?

Jill at Loughnaquilla (3039 feet) in the Wicklow mountains - Ireland 1991.

Chapter Twelve

To end of 20th Century – 1993 to 2000

Marriage to Jill – 6 March 1993; As I write this December 2018, we have just celebrated our silver wedding anniversary of 25 years! Last March 2018, the weather was terrible and only about 66% of people invited were able to get here. The 'Beast from the East' arrived but enough of that; more about it at the time. The marriage took place on a Friday at Hinckley Registrar's Office (see photographs) with family and a few immediate friends in attendance. We later retired back to number 47 for a late lunch/early tea meal, cooked and prepared by Ian – Rosie Barr's husband and very good it was too! Later, Jill and I went to spend the night at a bed and breakfast in a farm just outside Lutterworth!

It was then that I found myself suffering from gastroenteritis, contracted on my stag night at the Bluebell from Paul who sneezed over me! I stayed in bed as long as possible on the reception day, Saturday, and finally feeling awful, I staggered to Manorfield School to meet 180-odd, invited guests (see video by Nigel Davies). It all went well, John Irwin called it the wedding of the year. John Easten supplied the jazz band and brother Mick did a marvellous speech (Ref video) which drew much laughter from the audience and eventually Jill and I retired to spend the night at Hilda's home. The following day, I managed to survive a midday session at the Bluebell with the remaining guests before surrendering to my bed for three days or so. During this time, Roly our cat from Flitton died – killed on the main road, as I had feared and was buried with due honours performed by Mick and Jill, on the Green Lane behind the workshop, leaving one kitten who went to Jill's nursing friend, Jeanette, and was named Henry!

Almost all of the holidays mentioned in the text are photographically recorded and catalogued with places, people and dates, plus those included in this book. The following week Jill and I set forth on what could vaguely be called a honeymoon to Church Stretton, and over the Long Mynd past Cader Idris to Winsford Cheshire where Mick's daughter Caroline was staying and sharing rooms with Alan, Mary, and Holly who had just been born! They were, I believe trying to make a go of it to see if marriage was possible? No comment!

Whilst we were in North Wales, we went to Gresford where my mother spent her early years at their home called Laurel Bank, before she married. An endeavour was made locally to trace her brother Jim, who we were led to understand was the local GP for many years, following his father's practise. But unfortunately, he had died only a couple of years or so before we arrived. It was a shame in some ways because he married and had children who would of course be first cousins like Jill, Christine, and Martin Rush and John and Bernard Collett, all on my father's side. Just after this, my father died at the age of ninety-three-and-a-half on 6 of April 1993 and thereafter I spent some weeks going backwards and forwards to Ingleby Way, bringing furniture etc to Stanton; filling number 45 up, which had just been renovated after the tenants, the Copson's, had left it in a shambles!

We used Shelly's trailer and Paul's help to achieve this, some of the items also went to Phyllis in Wellingborough but she did not have much space for it. Prior to this, during the renovation of number 45 we had a bath and washbasin etc fitted with the idea of me looking after Dad (Jill was working full-time) but unfortunately it came to nought as he died about a week before it was finished! At the same time, I was improving numbers 47 and 49. Putting on new/second-hand doors and cupboards etc, redecorating and building bookshelves; fireplaces in the living room, front bedroom of number 49 and front bedroom in number 45 known as 'Nancy's!

As all this was taking place, my uncle John (Mother's brother) died, he was in his early 90s also. As he lived in a care home in

Herne Bay Kent, the funeral was naturally in that area. So, we went down and stayed with the Lanes, attended the funeral with Phyllis, John and Rosemary Collett, plus one or two members of the home staff. His effects were minimal as I think I have explained before! Our second holiday this year, May 1993, involved a trip to the Black Mountains – the Longtown area with the Lanes, Mick, Dorothy and Jack. This came about through Jack, who some years previously had had a holiday at Michaelchurch Escley. On a farm called 'Tycarrodog', Where he had been really looked after, good food etc. One amusing incident occurred one evening when it became obvious that the old couple were very religious and expected us to pray and sing, I think it must have been a Sunday night, so we four, and Jack excused ourselves, retiring to bed leaving Mick and Dot to do the honours!

We toured the area visiting Llanthony Abbey, Hay Bluffs, Offas Dyke and Craswall etc. This later led on to Jill and I visiting the area on several occasions, staying at a farm in Michaelchurch Escley for Bed and Breakfast with the lady we called Mrs Escley. She was lovely and always wanted to know what was going on in the outside world. i.e. London. Meanwhile, while as all this was going on, I was visiting the Labour Exchange (Job Centre) as I was technically 'unemployed' and in order to claim the dole, about £50 pounds per week at that time, I had to be seen to be actively looking for work until I reached the age of 60 when I would go onto 'auto-credits', i.e. they would pay my stamps etc until 65 when I would be eligible for the old age pension!

So, I started looking in the papers etc for suitable positions commensurate with my qualifications and experience etc. Needless to say, there was not much on offer for a man of my years, 58! I was not disappointed however, as I had no intention of working again; remember Jill was still in full-time employment and I had my own workshop and new engines to build, with Fred. Funnily enough, I found a branch of Howard Delta, the agency I had been working for at RAE in Burbage. The Job Centre in Hinckley I was dealing with knew nothing of their existence and so my standing with them rose and in addition to my application letters etc all went well until 1994 when 60 arrived.

Our third holiday took place with a visit to Diddlebury in the Clee Hill area of Shropshire. Jill had previously been here with the children many years before and had a desire to return. We had a stroke of luck when we arrived late afternoon in Diddlebury, we saw a hand painted Bed and Breakfast sign hanging on a lamp post beside a little lane which we followed until we arrived at the house! (Ref photographs 1993.) There was a man mowing the lawn and he turned out to be the owner of this huge place and after negotiations said we could have the whole of the top floor of which we had a choice of about six bedrooms!

The weather was extremely hot, and it was useful to be able to spread out. He cooked us breakfast every morning and sometimes an evening meal accompanied by himself and girlfriend. After some discussion it turned out that he ran a 'borstal type' school for maladjusted boys and knew of Ray, Jill's first husband who worked in a similar line. We got the impression that he was in financial difficulties and he was trying to sell the house he even offered it to us, as at that time we had not definitely decided to stay in Stanton. From here we returned home via Bridgnorth where Jill purchased our antique brass bedstead, now installed in the only rear bedroom we have. On another occasion we went to mid-Wales and climbed Plynilimon the highest peak in mid-Wales. it's about 2,500 feet, not a difficult climb, just a slog. Jill also came up with some difficulty as she was feeling unwell.

In September, we visited Yorkshire staying in Ilkley for three days, visiting Embsay railway, Bolton Abbey and Rombolds Moor. The highlight of the holiday was a visit to Otley to see David Barry, my old school chum; we first met at Sandals Road School situated at the bottom of Baildon Bank when we were about eight years old, in 1942!

Leaving Jill behind to talk with his wife Pearline, David and I decided to start from his old house in Baildon and do the same walk that we used to do with his father (mentioned in my previous memoirs of Baildon days). we took the back path and revisited Crook's Farm, Hope Hill, The Glen; viewing Saltaire and Shipley from the cliff path as we returned towards Baildon, calling in at the quarry, where we split up temporarily as I decided to climb up to Granny Hitchiner's bungalow, as we used to do! We then visited the old school and ended up at Tong Park; later retiring to the Bay Horse pub for a well-earned drink. (ref photos of all this).

To round off the Holidays for this year, we visited Devon and Cornwall with Lyn and John, staying at Widecombe-in-the-Moor, in October. We visited Trebarwith Strand, Polseath, Port Isaac – going to Polperro in Cornwall, where we walked the cliff path to Looe and saw Looe Island which the two ladies from Epsom purchased many years before; writing their famous book *We Bought an Island*, of which we have a copy! Also covered was the Lizard Peninsula where we also walked the cliff path to Kynance Cove and back; following this effort with a trip to climb Brown Willy, the highest point in Cornwall, not for the first time for John and me of course as we did it many times on the Three Tors motorcycle run!

Whilst all this was going on, Jill's father, suffering from dementia, was removed from his bungalow and taken to Colton to be looked after by George and Marion. He remained with them for about three to four years. Thus, ended in 1993!

Stopped writing between January 2019 and April 2019 as Jill died on 7 February 2019 and I have only just got back to it. Now 10 April 2019.

To continue with 1994; the sleeve valve 500cc four stroke engine was now finished and during the intervening period of 1989 to 1994, while such things as having to take control of my fathers' affairs, marital problems with my first wife, redundancy and moving house from Flitton to Stanton, plus remarriage, were taking place, the E2 was installed in its motorcycle frame, donated by Steve Linsdell. (Ref photographs and the book of the WC-E2).

Early in the year, January I think, I trailered it up to the council offices in Leicester, to be officially registered. The examination was cursory to say the least! Not technical in anyway; all the chap was interested in was whether the numbers stamped on the engine and frame, corresponded with what I had told him in my application letter and the receipts from various suppliers were authentic! i.e. not stolen. The machine was passed, and I was issued with a 'Q plate' as it was a special bike, like a kit car, comprising of homemade engine and frame, MZ mudguards head and tail lamps, Norton forks, gearbox and wheel, Royal Enfield rear wheel, Seeley fuel tank and Yamaha alternator! A real jumble: number Q223 HRY. I like to think the HRY stands for Harry Ricardo, the father of the sleeve valve engine.

After this, much riding took place over several years including four Three Tors events in 95, 96, 97 and 98 with this engine, a total of 5,808 miles being covered and a similar distance of 5,223 miles also with the two stroke 500cc sleeve valve engines, making a total of 11,031 miles ridden on the machine! over a period of 16 years from 1994 to 2010. With sleeve valve engines, all homemade by Fred and myself with much help from friends! Approximately 1,000 miles of test running etc was done around Stanton in 1994 and several incidents spring to mind.

There were two trips to Bedfordshire, the first to visit the Vauxhall UFO group run by Dereck Keep, and the second to pick up two Amal carburettors for fitment to the machine. They worked reasonably well. It was on this second trip, accompanied by Jill's brother Ian on his V-Twin Honda, that I ran out of petrol on the way home up the A5 and he had to find a garage and fuel to get me home, how embarrassing that was! On a further test run the alternator drive chain broke near Daventry when the grub screw in the sprocket worked loose and broke the chain spring link. I rode the bike home, after disconnecting the rear brake light switch, using only enough current for the ignition.

Also 1994 – I had a 60th birthday party at Stanton, attended by John Urwin, Dave Tebbut, Bob Durrant, Groococks, Mick, Dot and neighbours etc. Work also started on our garden room which was a new building constructed by the Groococks' to replace the old outhouses that also included an outside toilet; used for many years by John and Nancy Copson, Jill's tenants. (Ref photos). The new building was extended and joined onto the existing house number 45 Long Street. the back door now being replaced by an all glass one to let in as much light as possible, and extended further down the garden to the limit of the old wooden lean-to; presumably put up by John Copson.

Two other items of interest occurred this year, Iris started to paint with Jack's tutelage and completed some good drawings of plants/flowers etc for the Silsoe Agricultural College; and Mick, another of Jill's brothers, sold his butcher shop and business to a John Kilby and became – wait for it – a mental health nurse at Carlton Hayes asylum, at the age of 60! He was eventually transferred to Glenfield for another 14 years, retiring at 74 years of age not bad for an 'old un'.

About a month later in February 1994, Jill and I decided to visit the Preseli Mountain area of South Wales as neither of us had any previous experience of this mountain range. It proved to be very interesting; plenty of climbing; sites of archaeological interest, lovely coastline, all now in Pembrokeshire National Park; particularly spectacular was Carningli, a rock-strewn mountain situated behind Fiona's. Fiona's was our bed and breakfast for a week or so, we found it almost by accident. In those heady days we never booked any accommodation in advance, we just visited the area and took potluck! We saw the sign on the gatepost and that was that. We usually took a tent, just in case. Later in the year we returned with John and Lyn when touring Wales, visiting Aberystwyth where they broke the landlady's bed! She was not best pleased, but she was a miserable old thing, religious, all she wanted to do was get to church – Sunday morning, so no time for breakfast.

Our next trip was a big one – Orkney and Shetland and it came about via the Lane's. We had decided to go to the Outer Hebrides when literally without warning the Lanes arrived on our doorstep! They announced that they were on their way to Orkney and possibly the Shetland Isles, would we like to join them? Why not? No skin off our nose, nothing booked as usual, let's go! So, we all jumped into our blue petrol Renault and set off for Scotland. Our first stop was some 400 miles north of Stanton at the Allargue Arms Hotel, Cock Bridge on the A939 for a Bed and Breakfast. The views of the snow-capped Cairngorms were superb (ref photos)

By the second day we had reached Scrabster, calling in at John O'Groats on the way. On our way to John O'Groats on the A99 we called in at Wick, and John showed us where he had broken down on the Norton rally in 1990, where he had to rebuild the CSI to continue home as previously mentioned! We had a peaceful crossing to Orkney and arrived at Stromness in the early evening, during the crossing we saw the 'Old Man of Hoy' (ref photos). Now came a problem, we of course, had not booked any accommodation as usual and we found none in Stromness, so we headed out into the countryside going north! Eventually, just as daylight was fading, we arrived near Marwick and saw a bed and breakfast sign. It turned out to be a farm and the owner, a Mr Clouston, had a woodworking business and fully equipped workshop, very impressive!

During our time here, about a week, we visited various sites of interest, I'll mention them, but they are not necessarily in order. On the mainland, we went to Kirkwall and its Abbey, this is the capital of Orkney and a remarkably interesting and old centre it is. Skara Brae – Prehistoric Cairn, which was revealed by the sea during a huge storm many years before. Kitchener's Monument, just up the road from our bed and breakfast, his ship sank just off the west coast shore here with the loss of many hands. Birsay, again close by, we used to visit a pub here for lunch and visited the Orkney brewery to try the local ales! Going further afield, we visited South Ronaldsay Peninsula, to the Tomb of Eagles (I have a book on this subject) and the Italian Chapel where Italian prisoners of war were stationed and who made some very interesting artefacts. Very clever!

Turning now to the islands of Orkney, we went first to Hoy which is the most mountainous and climbed the highest, Ward Hill, at about 1,600 feet, it was a tough climb due to the heat, an extremely hot day indeed. Lyn lost her binoculars near the summit – unfortunate! We also visited Betty Carrigall's grave, which is at the roadside on the way toward Hill. It appears that she was a 'fallen woman' and no church would bury her so out she went, not much has changed! Finally, a visit to Rousay Island in the north, where we hired bicycles and pedalled around the coast road in glorious sunshine (Ref photos).

So, onto Shetland, the seven-hour crossing was superb, beautiful weather all the way, like a Millpond. Upon arrival at Lerwick, we discovered that some of the large sailing ships were in the harbour and provided an impressive sight. This time we had booked a Bed and Breakfast in advance, again arriving in the evening and it was located at Tingwall in the centre of Zetland which is the mainland, at the head of Loch Tingwall.

The B&B was an old Manse and the place where the old parliament used to convene. On its right is Scalloway the ancient capital before Lerwick. It was run by a young mother who had a four-year-old daughter whom I used to read to in the evenings before she went to bed. Talking of reading, because it was nearly Midsummer, the sun only went below the horizon for about five hours; I found I could read a book outdoors at midnight! Amazing. The following morning, having reviewed the menu, John and I decided against the full English breakfast, so we asked if it was possible to have boiled eggs. When it arrived, we were presented with the full English _and_ boiled eggs, marvellous! From there we headed west towards Bixter, past a few 'side of the road' scrap yards, mainly old farm equipment, some cars!

Bridge of Walls, Muness and Sandness provided stunning seascapes with views of the Isle of Foula, far out in the Atlantic. It was the scene of a film called *The Edge of the World* made in 1937, starring John Laurie, who was later to achieve fame as 'Fraser' in *Dad's Army* and with a title like that, you can see why they chose it. Looping North, we headed for Ronas Hill at 1,500 feet, slightly lower than Ward Hill on Orkney. Unfortunately, we were unable to climb it as the going was extremely rocky and the presence of a military base nearby was discouraging. A day or so later we were at West Burra, South of Lerwick where I noticed a little boat coming in on the tide and I waded out to get it. I still have it, as it comprises a shaped wooden hull with half a tin lid stuck in it to form a 'keel' at the rear, with a wooden stick enclosing a piece of plastic to form a sail at the front (ref photos) And/or the real thing! There seems to be no message on it, no identification so I wonder where and how far it had come? Weird.

A trip far south to Sumburgh Head was accomplished and many birds of all types seemed to gather here including many puffin's, to Lyn's delight. We said goodbye to the people at the Manse after about three days and headed north for the ferry across to Yell. E84 JRK Was still going well in spite of a knock from the front CV joints!

We were now back to our old tricks of not booking bed and breakfasts in advance, so we headed for Burravoe and came across a sign; so, booked in but failed to look round as we were keen to sightsee; a disaster. It was the worst B&B I think we have ever stayed in, in all the holidays put together! It was in the southeast of the island at a place called Burra Voe and the proprietor was a Mrs Williamson. For the evening meal we decided on lamb and vegetables but when it turned up it was mutton and greasy vegetables all running in fat! Mountains of it, the son of hers had cut off slice after slice, he seemed to have no idea of portion sizes, Needless to say we ate very little. Next to bed; well the rooms were damp with the wallpaper falling off the walls and gaps around the window frames, plenty of fresh air! Breakfast the following morning was no better, John and I decided on the full English – mistake. The girls, having more sense, went for boiled eggs, ours was just a greasy mess, everything running in fat. However, we left as soon as possible and headed north via Gutcher and Gloup to arrive at the ferry to Unst.

It is a particularly bleak place with virtually no trees, just open moorland. We had a lunch stop at Murness Castle in the south of the island and then headed north as far as we could go via Saxa Vord which is the highest point at 910 feet, not quite a mountain, which we didn't climb. Finally, we arrived at the most northerly post office in the British Isles, just south of the Noup, which is the most northerly point in Shetland, (ref photos.)

During our time here we went for a walk across the moors and were attacked by Bonxies which are a great Skua, they dive bomb you to protect their young at this time of year, they are a ground nesting bird and very formidable. We returned to the Scottish mainland via Orkney and spent another the night in Stromness before returning to Scrabster. From here we went west along the north coast of Scotland to Tongue for another bed and breakfast, deciding to explore the area, visiting the Smoo Caves on the coast and returning inland to climb Ben Hope. Ben Hope is the most northerly 'Munro' at 3,040 feet. Ben Lomond is the most southerly – just thought you would like to know!

We climbed it virtually from sea level, no cheating! Only three of us, as Lyn wasn't feeling up to it. We ran into snow still lying around at this time of year, but it caused us no problems. We toured the area going south generally, visiting Altnaharra, Ben Loyal, Queen of the Scottish mountains with its five summits. Foinaven, Ben Arkle etc passing through the Glen Canisp Forest which includes such spectacular peaks as Suilven and Stac Polly, which I subsequently climbed on a later holiday.

Making a diversion to the Applecross Peninsula, climbing the spectacular Alpine-type pass (pass of the cattle) Old Drove Road, over Squarr 'A' Chaorrachain at 2,515 feet and into Applecross village for lunch which looks like beer and fish and chips from the photo! Superb views in all directions as the weather was still good. Arriving at Lochcarron near the Kyle of Lochalsh, Skye for our final B&B in Scotland, almost all of them without pre-booking, so much for your WWW! From here the following day we struck inland towards Invergarry on Loch Ness, thence past Ben Nevis etc and south into Glencoe at the King's House Hotel and finally back home. via Yorkshire for our final B&B at Muker near the Buttertubs pass to Hawes. Touring through Yorkshire, we met a motorcycle rally in Kettlewell who were doing a Land's End to John O'Groats run, there was of course no holding John back! We bought some books and had lunch here. Exhausted, it tires me out writing about it all, we must have had some energy I was 60 in 1994!!

Our final holiday for this year was in October using the Lanes car; ours had done the previous slog. The same team set off after meeting up here at Stanton, heading north for Yorkshire where our first port of call was Embsay, Steam railway station, first visited last year by Jill and myself. From here we went to Todmorden, previous practise of the notorious Doctor Shipman and now home to the Brodrick's. When they left Kent, as I think I've mentioned before, they used the money from their house sale to buy a farm in Yorkshire, and here it was! It was the first time John and Lyn had been there since they moved.

On through Swaledale, Deepdale to arrive at Tan Hill in the mist, not surprising as it is the highest pub in England, nearly 2,000 feet altitude. I think we stayed for B&B at Mukor as before, but I'm a bit shaky on these overnight stops for this holiday! nothing however was pre-booked, as apart from calling on the Broderick's, everything else was on spec! Changing our minds as and when, great! no deadlines to meet etc, unheard of today's slaves to the mobile phone: it's called Freedom!

A democratic decision now directed us to Hadrian's Wall and into southern Scotland, to Moffat and east along the A708 towards the famous waterfall called the Grey Mares Tail. We found a nice

B&B near Mountbenger crossroads, where we stayed for three days in order to do some walking/climbing to avoid too much time in the car! It was called Gordon Arms Hotel and was run by a man called 'Harry Dum Dee Dum'. Not his real name of course, he was christened this by John Lane who said, "hadn't we notice all the time he is rushing about the premises, he is saying, 'Dum Dee, Dum Dee Dum'."

We climbed up alongside the Grey Mares Tail and up into the Tweedsmuir hills towards Broad Law, Dollar Law and Black Law which are the highest points around here, but not quite the highest in southern Scotland as Merrik further west has that distinction. A visit to Hawick to see the Jimmie Guthrie memorial rounded off our time in this area and we moved westwards across Duchal Moor, where we attempted to do some kite flying but the wind wasn't playing ball, although it wasn't all failure.

Jill and I climbed Creuch Hill at 1,500 feet the highest point on the moor with great views of Arran in the distance. The weather was great and that settled it, we had to go to Arran. Sailing from Ardrossan to Brodick takes about an hour and a half; millpond crossing, and we set off north to end up at Lochranza where we booked into the local hotel for a few days. I remember particularly the breakfasts; we had the biggest and tastiest kippers I have ever eaten! We toured all over the island and very beautiful it was. It has been described as a miniature Scotland with the lowlands to the south and the Highlands in the north! We had to climb the highest point of course, so we tackled Goatfell at 2,840 feet and all reached the top after a fairly rugged climb. (Ref photos.) From here, it was home again to round off 1994. I don't know how we found enough time and energy to do it all; good job most of us had retired by then, except Jill of course but then she always had an excess of energy! Hence two husbands, two allotments, big garden and house and idle second husband!

1995 March. As reported before, about 1,000 miles of road testing had been done on the 500cc four stroke E2 Sleeve valve machine on a 'Q' plate registration. An article in the *Classic Motorcycle* magazine by Brian Wooley was published and almost immediately I was contacted by two other special builders. One, a Doctor Will Hutton from Horsforth near Leeds and Hedley Cox from San Antonio, USA. Correspondence ensued and a visit to see Doc Hutton was arranged and Jill and I went to see his engine, it wasn't a sleeve valve but he had modified a 500cc engine so that it inhaled its charge into the crankcase via a reed valve pack on the up stroke of the piston and delivered it up a pipe into the normal inlet valve on the downstroke! A similar principle to Willy Wilshire's 'Willhelmthing' Based on a Triumph twin engine in the 1950s! He had done several 100 miles on it and claimed a power improvement due to the supercharged effect. I was unable to ride it myself due to a problem with the gearbox mounting, (ref letter.) A very nice man who had been a GP but had to retire early due to ill health; said he much preferred engineering anyway!

The other correspondent Hedley Cox, I never met but we kept up writing to each other until 2015, 20 years! (Ref correspondence.) Most of the time, we commiserated with one another trying to solve each other's problems and encouraging support. He had a similar career to me, automotive and aerospace, finally going off to the USA after he fell out with Bertie Goodman of Velocette, we've heard that one before! He had concentrated all of his efforts on the sleeve valve four stroke, having made a 250cc single cylinder in the 1950s and a 500cc V-twin based on a Honda bottom half and possibly one or two others I haven't heard of, or forgotten.

He seems to have had much trouble with getting his sleeves round! Mainly, I think because he hasn't recourse to accurate grinding as far as I can tell from his letters. His workshop looks reasonably comprehensive much like mine; but old fashioned, much like mine! In 2009, he wrote a book called *From the Race Shop Floor* which was an autobiography, incorporating the final chapters on his sleeve valve activities. This was followed in 2015 by *A Guide to Motorcycle Racing* which I have not had the pleasure of reading due to its high expense! Altogether, a very interesting man.

It was earlier this year, February, that Lyn Lane had a brain haemorrhage similar to the one that killed Ray, Jill's first husband, but she survived, as it happened one afternoon at work and a colleague recognised the symptoms, rushed her straight to the adjacent hospital immediately, thus saving her life! However notwithstanding, she accompanied us; that is John, Jill, Jacqueline, Alan and me to Jura in May, in their car with John driving. We rented a place at Ardfernal on the east coast of Jura in a very remote setting with electricity but no water, it had to be obtained from the adjacent well in buckets provided. As soon as we arrived and settled in there was a problem! No Jacqueline and Alan?

They were coming separately in a hired car and were getting a later crossing, but after a six hour or more delay, we were beginning to worry, remember no mobile phones then! Eventually towards evening, they appeared, and it transpired that they had forgotten to bring the address with them; unbelievable! They had been rushing up and down the island, asking various people if anybody knew if some English holidaymakers were in the vicinity; we were a little off the beaten track.

We had a week here and spent the time walking as much as possible; we walked across the middle of the island to Loch Tarbet in the west and re walked our 1989 walk, we had done with Tom and Sylvia to Ardlussa, looking once again at George Orwell's house, as before. We also climbed part way up the Paps except for Alan who went to the top of one of them into the mist and it was during this expedition that Jill nearly got bitten by a snake.

She was running downhill towards me and as she bounded over some Bracken, this Adder reared up and struck at her, but she was going so fast that it missed! Obviously, it had been taken by surprise as normally they are reclusive creatures. It was during these walks that we encountered ticks! These creatures are found on sheep, deer and other grazing animals all over Great Britain, but it was the first and only time in all of the walking we did that we experienced any problems. There are more than 6,000 deer on Jura, So I suppose it was inevitable.

Several of us had them on our legs but the most serious was the one near to Jill's eye which with care was removed. If left unattended they can lead onto Lyme disease. Nasty! Interestingly, we had another pest, if you like, not so dangerous, more interesting, a mouse. He used to live behind the skirting board in the living room and used to appear nightly, running across the floor which proved most entertaining, better than television, although we didn't have this on holiday; so, reading and sleeping was the order of the day/night.

Although as I said we had electricity and water, we had no heating in the living room except a fireplace and chimney. So, what to burn in the evening. We had noticed on our travels, at the side of the road in places, were piled up mounds of peat. So surreptitiously, at night we motored out and filled the boot of the car up. Job done! Finally, a visit to the whisky distillery at Craighouse and the only pub on the island for a well-earned pint.

Now we had to say goodbye to Jacqueline and Alan as their week was now over and we were on our way to Islay via the Feolin ferry for our second week at digs in Bruichladdich, There's a mouthful! Near Port Askaig. Later, we visited Bowmore, the capital and a wrecked ship on the beach close to the whisky distillery Bunnahubhain; one of six distilleries on Isla. A trip to the Mull of Oa, where on a clear day you can see Ireland was interesting for two reasons:

1. The American monument on the cliffs is dedicated to the Ilay men, by President Wilson in 1918 who gave their lives in World War One.
2. A jellyfish with a skull like face inside it (ref photo.)

Further along the south coast, we came across the distillery at Lagavulin and its ancient castle, followed by a visit to Kildalton's ruined church, where we saw a peculiar type of bird called a Treecreeper. Finally ending up at Ardtalla, the end of the southern road around the island. We had a picnic here and with our usual 'gusto' decided to climb Islay's highest point, Beinne Bheigeir at 1,600 feet. Lyn declined for obvious reasons and John, Jill and I went to the summit to record excellent views. The rest of the island is mainly moorland especially in the north where roads are few and far between, but all views are stunning and wildlife aplenty, including many seals around the coast at places like Portnahaven and Port Ellen etc.

Upon our return, I received a phone call from Clive Heasman during which he mentioned that Gordon Cornell lived in Coventry and suggested I phone him to talk about engines!

"Gordon Cornell!" I said, "who's he?"

"You probably remember him as, 'Pepper'," he said, then the penny dropped – an old aero modelling friend from Wallington days as I have mentioned before. The net result of this was a reaffirmation of our friendship and the start of a computer programme called ICE – 'Internal Combustion Engine' Which eventually, he was to sell all around the world on the worldwide web.

His career was similar to mine in engineering; after getting his qualifications at Croydon Tech, he worked for Electronic Developments (ED) in Kingston, developing their range of model aeroplane diesel and glow plug engines followed by a period in the motor industry with Commer cars in which he rose to executive status before turning to producing his own range of high-performance compression ignition and glow plug engines sold under the name of 'Dynamic'. These have been built on the development work he did during his early years and are among the 'leaders in the field of high-performance model aero engines', quote from *Model Engineer* magazine.

Now ICE. began as INDIC – 'Indicator Diagram for Internal Combustion Engines' And was started by me at Vauxhall's in the

1970s, as previously mentioned, we decided to scrap all the previous four stroke analysis and concentrate the programme on two stroke engines only, also converting from Fortran to Visual Basic. We got together several times a week for about four months and the preliminary program was running by October of this year and several articles were published in the *Model Engineer* popularising it, such that sales commenced via the WWW which also included a comprehensive user manual, ALL retailing for about £45.00.

All this was done by Gordon, he purchased several printers etc and set these running all day to keep up with demand from all over the world! Amazing, it took off straight away, I took no revenue from any of this, not wishing to get involved with possible tax problems! The program has been updated several times since then and I believe is still selling, although at a reduced rate; at the time I am writing this, i.e. August 2019! 24 years on! Neither of us visualised that. We also learned a lot from the experience and our knowledge and understanding of the two-stroke engine increased considerably.

Meanwhile, the 500cc sleeve valve four stroke E2 in Steve's handmade frame had been receiving further development and was now deemed reliable enough to go on the Three Tors run around Devon and Cornwall. At the last-minute Ian – Jill's youngest brother decided to come along riding his 700cc Vee twin Honda which only had a small fuel tank necessitating frequent stops for petrol! We met up with John, Derek, Jack and Steve at Clive Heasman's house in Devises And rode on to stay with Geoff Brown at Goodleigh, as usual. We completed the Three Tors successfully. Ian, with the usual Vernon energy actually ran up several of them, maybe others as well! We visited Keith Bramwell's abode to be made very welcome. He and Jack got on well together both being artists in their respective doctrines.

My sleeve valve was very fuel efficient, i.e. about 80+ to the gallon, same as Johns CSI but the oil consumption, not brilliant but comparable with Steve's 'Black Pig', i.e. 1200cc side valve 1938, Royal Enfield. The total miles covered without any major problems was around 750 to 800 miles approximately. The machine now had completed 2,400 miles in total and a major strip down was done, (ref book E2) and several small points attended to.

Also, around this time, Jill and I passed through Chelsea and discovered, to my horror that my old college was no more! All dereliction: Sydney Street completely changed, although Kings Road seemed untouched. So, on the strength of this, we visited Redhill Aerodrome on our way down to the Lane's, to discover also that although the hangers were still in place there was no sign of my college. Subsequent inquiries revealed that the entire institution had moved to Shoreham on the Sussex coast and was now called 'Northbrook College of Further and Higher Education', sounds posh!

So, I wrote to them, explaining that I studied there from 1956 to 1958, obtaining a first-class diploma in aeronautical engineering, which has stood me in good stead for the rest of my life and asking if they had any relationship with my former college and whether there was an 'Old boys' organisation or suchlike. Their reply was cordial; explaining that Chelsea had been absorbed into a Federation of colleges in 1986 and become a Department of Northbrook College. The changes that have taken place are too many to list but if you would like to visit, they would be happy to show me around and bring me up to date. Needless to say, I have not followed it up. Typical!

Our next major holiday this year was to Ireland with Lyn and John again. This time using our car, a petrol driven Renault G981 ALT; Jill and me driving. According to John, after meeting here (Stoney Stanton) We went via Holyhead to Belfast and thence across Northern Ireland to County Donegal to arrive at an 'old world thatched cottage'. This had been booked in advance by Lyn and proved to be extremely primitive but superbly different. It had been modernised to the extent of running water and electricity but little else that people take for granted these days! Two tiny little bedrooms, one tiny little drawing room and one tiny little kitchen, plus loo! Lovely. It was situated just outside a small town/ village of Ardara which was the place that Martin Fenny and his father visited for a holiday in the 1950s; it must have been bloody remote then.

From here we went on several forays to explore County Donegal, after first examining the local area i.e. the beach at Ardara where there was some stunning caves and an impressive waterfall, followed by a trip up the 'pass that wasn't a pass' which took us up to a view of a valley that was very beautiful but went no further. From here to Maghera with superb uncrowded beach, caves, and a further waterfall. On to Glencolumbkille, along the coast road past Sleive Tooey at 1,458 feet, finally turning south, still following the coast road to Killybegs, a small fishing port that I visited with Iris in 1968.

Portnoo stands out as a spectacular area with massive cliffs, rock formations and incredible seascapes, well worth a visit, all on your own doorstep, so to speak no need to go abroad! Central to County Donegal includes such places as Glenties, Flintown, Letter Kearney and a look at Lough Garton; all were included before undertaking our first real project – the ascent of Errigal, Donegal's tallest and most spectacular mountain at 2,466 feet! A perfect cone!

This glorious sight was visible from the front of our cottage looking north (see Photo) and I used to spend considerable time peering at it through my binoculars. So, on the first available fine day we set off, heading for Gweedore and Rosses Point, thence to the foot of Errigal itself. Lyn, still recovering from her aneurysm, declined and it was John, Jill and myself who set off with great enthusiasm for the summit, which we reached about an hour afterwards. Errigal has two summits about 150 yards apart, Jill and I did both but John being not sure of foot, and with steep drops on both sides of the ridge, took the wiser option and stayed where he was on summit number one. The views of Slieve Snacht 2,240 foot, Muckish 2,197 foot, and the poisoned Glen were 'awesome', horrible modern term, but we all returned safely. We then went to the Poisoned Glen, explained in local guidebooks, and had a well-deserved tea and cakes in the local heritage centre, i.e. village Hall, finally having a look around the local church.

Later in the week we toured around the Atlantic Drive visiting Carrickart and the bloody foreland which of course, John and I had done previously in the 1960s. Returning via the West Coast Road through Dungloe etc, we found an interesting, ruined village near the Tory Sound, (ref photo.) I wondered if it was possibly the village we camped in 1961 with Anne and Richard Symes, when John and I 'did' as many pubs as possible that night? Possibly. It must they pretty close to this area! Another first for us was a trip to the Irishowen Peninsula which leads to Malin Head the most northerly point of all Ireland and it lies in the Republic.

Avoiding Londonderry, we travelled up the east side overlooking Loch Foyle and arrived at Malin Head which apart from the very smart lighthouse and sea views, is not particularly special, we've probably been spoilt with all the previous entertainment. Returning down the west coast through the gap of Manmore was I think, the most spectacular. This was now our last look at County Donegal as we turned our car southwards and headed for Sligo via Ballyshannon, Bundoran and Lough Melvin, arriving later at Ben Bulben 1,722 feet, a very spectacular mountain with a sheer extended face, facing west but gently rising on its eastern flank. We spent some time here where there was an impressive waterfall as well as the poet Keats grave.

From here we turned west and headed for Owenbrin, crossing the Slieve Gamph or Ox Mountains and thence into County Mayo to view Nephin and the Nephin Beg Range alongside which the Owenmore River flows. We booked a night or two at Foxford and using this as a base, we ventured through the Nephin's and onto

Achill island. Foxford is almost on the shore of Lough Conn and is noted for its swans, hundreds of them all dotted over the lake. Jack would have loved to see them! And did so on a later holiday. On Achill, we visited the beach on which we had camped in 1961 with Richard and Anne and visited a deserted village taking shots of Clew Bay with Clare Island in the middle of it. Then on towards Keel with great views of Croaghaun, 2,192 feet, although I don't think I've had a clear view of it in all the photos I have taken over the years, before or since! Extraordinary, it's always in cloud!

Leaving Foxford, we continued south passing through Newport and Westport, turning westward along the coast road. Past Crough Patrick 2,510 feet, Irelands holy mountain, where the pilgrims climb, some bare foot, to once more traverse the Louisburgh to Leenane Road with all its glorious views; passed Mullrea, 2,688 feet an old extinct volcano and on to Cong. Here we stayed for three days in a hotel right in the centre opposite the quiet man cafe and Cohen's bar which was the final scene of the famous fight in *The Quiet Man* film! 1952.

We spent a lot of time exploring Cong and the surrounding area of beautiful Connemara in the successive days, such that I have decided to split it into two parts: -

Part One – Cong and its environs: – We stayed in Daxagoners Hotel in the centre, opposite The Quiet Man' coffee shop alongside the Pat Cohen bar; to the rear of us was the Abbey, good view from our bedroom windows and beyond that the road leading over the river, past the fight scene and towards Ashford Castle in its beautiful grounds.

Part Two – Joyce's country and surrounding area. we paid a visit to see Teresa's sister Joan, who lived somewhere near Tuam, I believe. She, her husband, and daughter were having a house built or an old house restored, either or!

During the ensuing conversations *The Quiet Man* film got mentioned and she said some of the action took place nearby. So, we visited the site of a river where John Wayne took off his shoes and ran across after Maureen O'Hara! Well, who wouldn't?!

We further visited the three main groups of mountains in Joyce's country, namely the Twelve Bens, Partry and Maumturk Mountains and their associated towns; Cliften, Nafooey and Laough Mask and Recess during which the weather became variable. Continuing south we entered County Clare by the coast road through Ballyvaughan to Blackhead and onto Slieve Elva 1,109 feet, highest point in Clare and onto the Burren.

For some time now it had been apparent that the Lane's camera was faulty in letting in extraneous light etc, such that we were now dependant on my (Iris's) Benchini using 120 film which was becoming more difficult to obtain. Only wedding photographers were using it, I believe! The Burren is an extraordinary place, (ref guidebooks) and well worth a visit, so much rock, yet still small plants seem to thrive; incredible! Moving on, we passed Doolin Point to get our first glimpse of the cliffs of Moher which we visited next. All except Jill had been here before, although Jill had been to Ireland before with Jack and myself, in 1991– we had not ventured this far north, confining ourselves to the far southwest. Interestingly the Aran Islands are visible from this coastline, Irishmore the largest, Irishmaan and Irishmeer of which, I believe, Irishmore is the only one inhabited.

Moving on yet again, we now arrived at Ennistimon, the scene of our previous 1962 visit. We managed to find a suitable bed and breakfast and then revisited the old railway station that John I had slept in, in 1962. The nearby water pumping tower is now the health centre and Mrs Carrigg's bar etc, we used to visit for tales of the 1920s, Black and Tans etc is closed. Finally, a visit to the river and successive waterfalls left us with a feeling that at least some things had not changed.

Turning east now, we headed for Dublin, crossing the centre of Ireland; a ruined house near Cloghan is the only record we have of this journey; it is possible I was running low on film, saving the remaining shots for the Wicklow mountains; before we ran into trouble near Dublin. Some little time before I had noticed that the engine was getting less responsive, sluggish on hills etc and eventually almost unable to run at all! Fortunately, we were on the outskirts of the city and managed to 'limp' to a garage. We had no AA or otherwise cover in those days; we just took some tools and hoped for the best!

They said it seemed like the carburettor (now extinct) and would take a few hours to fix, therefore we realised we would miss the afternoon boat, so leaving the car in their hands we took a bus, went to the docks and made arrangements to catch the early evening boat and spent some time sightseeing around the local dock area. all went well thereafter, we arrived in Holyhead in the evening, found a B&B overnight and took off the following morning down the A5 past Tryfan which I had climbed in 1959 and so, home to Stanton. This was the last time that Lyn and John together went to Ireland. Finally, to round off the year, Jill and I went down to Bedfordshire near Newport Pagnell to see Miles race on a grass track using his 250cc Kawasaki, he finished several races somewhere in the midfield if I remember correctly. (Ref photos.)

1996. This year we had what I believe was one of the earliest holidays ever – we left for Woodbridge, Suffolk in February with snow on the ground,) We were visiting Bob Durrant's cottage 'next the sea' to meet up with John and Lyn as usual. we spent a week in their cottage with Durrant and his parrot; yes, he had bought his pet parrot with him and spent a considerable amount of time cuddling it in the evenings; don't ask! A lot of the time, as usual, was spent touring around the area visiting places like Southwold, Aldeburgh and Chapel St Andrews, near Boyton where Bob Durrant found a gravestone with his name on it, Robert Durrant. (Ref photo.) Fortunately, not relevant, he was to live another 22 years!

Using the petrol engine Renault G981 ALT (we didn't go to a diesel until around 1997/98) we made another visit in March to Fiona's B&B in the Gwaun Valley, Preseli mountains area with Jack where we walked locally exercising Fiona's dog on the way to an interesting waterfall behind her house. Also visiting Newport and surrounding area where the snowdrops were plentiful and beautiful! We climbed Carningli again (without Jack) understandably as it is extremely rocky and apparently the second highest in the Prescelis! We also visited Margaret and Allan Marriott at Kittle in the Gower where great interest in painting was shown between Allan and Jack, to both their mutual advantage. Whilst we were here, we observed the Japanese discovered 'Hyundai' comet through binoculars, just visible to the naked eye, but extremely nebulous. On the home front, a change was taking place! Jill and I decided to try and extend our existing garages from two to three! There was a spare Patch of land, actually brick flooring, part of a row of cottages that stood there when Jill was growing up. She had turned the other two into garages when she took over her share of Abbots Yard in the early 1980s. But there was some doubt as to who owned this piece of ground.

Sandy's children, next door at 43 Long Street, played on it fairly frequently and she claims that it was on her deeds, which is reasonable knowing Neil the builder, who had constructed her house! So we decided to avoid any further trouble, to offer to purchase it and after a small amount of banter, agreed on £500, whence the Groocks moved in and for a further £2,000 the job was done and we were in the position of having, one car and three garages, whereas most people have one garage and three cars!

Our third holiday in May of this year was back to Ireland once again in ALT. Jack, Jill and myself sailed from Fishguard to Rosslare arriving late afternoon, motoring to Kilmore on the south coast for an overnight B&B in an old schoolhouse! The following morning, we headed west towards the Comeragh mountains just short of Clonmel and drove well into them from the southside, climbing up to a beautiful waterfall where the views seaward were stunning. We then moved further west into the Knockmealdown Mountains, which are more rounded, through the Vee gap to arrive at Ryan's pub in Clogheen for refreshments and to receive directions to our next B&B at Ballyboy House, a posh farmhouse hotel which was run by John and Breeda Moran.

It immediately Rang a Bell – 'Breeda' Dalton, the girl I used to go with circa 1958 and we were in the right area, Tipperary/Thurles where she used to live. Subsequent inquiries revealed that this Breeda was in fact her cousin and that she was married to a builder I think and lived in Tipperary about 15 miles away, but she would not give me the address! Very strange. we concluded that they were heavily Roman Catholic and did not want to stir up any trouble, neither did I, but it would have been nice and interesting to find out how she had got on in the intervening years – Jill didn't seem to mind.

Our ultimate destination, in case you were wondering, was a cottage we had rented for a week in the Glen of Aherlow on the northside of the Galtee Mountains. However, in order to arrive at the right time 4:00pm, we had to kill time and decided to visit various places on route starting with Clonmel and onto Cahir, where we visited Cahir Castle with its ancient mediaeval bridge, thence to Cashel and the famous rock/Abbey with its religious connotations, and finally Athersal Abbey at Golden. Note: a map of the area would at this point be useful! Thence to our digs. having collected the keys from the farmer/landlord we moved in, (ref photos); the place was situated on the northeast side of the Galtees near to the River Aherlow.

Extensive climbing and walking now commenced for Jill and myself with Jack concentrating on filling his sketchbook and some limited walking at level ground. The Glen of Aherlow is an extremely interesting place running east/west along the northside of the Galty mountains, it has steep quarries and the River Aherlow on the east side where our digs were, we spent a lot of time exploring these. Moving westwards is the high ground of Galtybeg and Galtymore at 3,018 feet, which of course we had to climb again; remember we had last climbed it in the mist with no views from the summit and a dangerous descent! Circa 1991.

So, choosing a fine day, we set off and slogged our way up Galtybeg and on to the summit of Galtymore Where the views in all directions were brilliant; being near the centre of Ireland, the views were mainly all land; sea to the south, just visible. Whilst up here we also climbed Cush Cois, a lower knob shaped peak with good views of our cottage to the northeast. Finally, on this range to the west, lies Templehill, a pyramid shaped mountain about 2,500 feet altitude which we climbed separately, leaving Jack sketching at the bottom. When we returned, he told us of a hare that had sat with him for many minutes, until he moved, apparently mesmerised unable to quite make him out! We did do some other things besides climbing/walking; one night Jill and I went to a 'gig' i.e. Irish music and a barn dance some miles away and stayed out all night, merry making, finally creeping in in the early hours so as not to wake Jack up!

We also visited the town of Tipperary; I remember dogs lying about in the High Street which looked partly unmade. Finally, a visit to a youth hostel on our way towards Michaelstown where Jill and I descended into the bowels of the earth on a guided tour of an extensive cave system under the mountains – very interesting. Jack did not come down as the descent involved using ropes on the vertical sections! So finally, our week came to an end and we set sail for pastures new. We were headed to County Mayo for our second week at a place known as Lettermaghera near Newport. On the way, we saw a substantial forest fire near the Stevefelin mountains, just before we went through Limerick and onto Ennis in County Clare, to eventually arrive at the Cliffs of Moher for my fourth time and Jack's first; naturally, he was mightily impressed.

From here we proceeded to Galway following the coast road and thence to Connemara where Jill and I did a quick climb of

Bengorm 2,303 feet to get great views across Clew Bay etc and then on through Westport and Newport to our accommodation close to Lough Feeagh. It was called Pike's Cottage and rented to us by a Mrs Boyle, or so we thought! The true situation will be revealed later with the arrival of Jacqueline! The views were magnificent, to the south, Clew Bay and Croagh Patrick, Ireland's Holy Mountain. to the east Buckoogh or the 'aerofoil' mountain as Jack called it, as it looked like the fin of an aircraft. To the north, open moorland and behind us to the west, the side of a mountain, name unknown.

Down the road/track were two of Jacks 'friends' a lovely black horse belonging to our landlady, so we supposed, and an extremely long horned ram. We were here for a couple of days before having to go and pick-up Jacqueline from Knock Airport during which time Jill and I climbed Claggan, a small 1,500-foot hill situated between Newport and Westport that gave us good views of Connemara to the south, Achill and Mullet northwest and Nephin and the Nephin Beg Range, due north.

We arrived at Knock airport at about 10:00am and returned with Jacqueline through the Ox mountains and Foxford, to see the swans, viewed the previous year with the Lanes, where we did some shopping etc and finally home. Trouble started a day or two after this when Jacqueline decided to have a bath and afterwards, we found that we had no water! So, a visit to Mrs Boyle's farm just down the road and a visit by her husband who disappeared up our garden to discover that our water tank was empty! He spent the next hour or so bringing water up from his farm to our place, refilling the tank. (ref photo.) When we inquired as to why he hadn't supplied enough knowing that visitors were due for a week, he replied that they were not the landlords but that it was owned by a reverend gentleman whose motto was 'every penny is a prisoner' i.e. mean, presumably he was on a water meter? Mystery solved. Jill and Jacqueline spent a lot of time together on various local walks and one of theirs is called 'coming home' and it is a scene painted by Jack, partly from a photograph which now hangs on our lounge wall, to be given to her and Richard eventually!

One of our first forays was to Achill Island and visit to the chemist in Keele for some sunburn lotion for Jacqueline of course – I don't think we had that much sun! Then on round the island finding the most magnificent fuchsias, a prehistoric cairn and a lonely grave dated 76, presumably 1876? Next, was a trip to Buckoogh Mountain (The Aerofoil Mount) which was clearly visible from our dwelling. It was one of the toughest climbs we have ever done, not that it was particularly high, say 1,500 feet, but nobody ever climbs it; no footpaths, just solid scrub and gorse, where you had to pick your feet up every step. Jack stayed at the bottom sketching as usual, sensible fellow. We made the summit and had great views of Clare Island, Clew Bay, Crough Patrick to the south and our cottage on the flank of Ben Gorm.

Nevin at 2,646 feet is the highest mountain in County Mayo, and of course we had to climb it. A photograph of our start was taken by Jack and he did a drawing of it which again hangs on our living room wall – to be given to Richard and Jacqueline. We attacked it head on as is my usual approach and this resulted in scrambling up a difficult rock-strewn gully, which Jacqueline said she was not coming down again! We arrived at the summit in glorious sunshine and breathtaking views, but it was very rock strewn, we came back another way to please Jacqueline and retired to the local pub for refreshments. During our time in there, we got into conversation with the locals who generally thought that we had come for the fishing and could not understand why anybody would want to climb mountains, it seemed that almost all of them had never climbed Nevin and no intention of doing so!

Jack became friendly with one of them called 'Desie' and took his photo outside the pub, later sending him a print, when he got home to Flitton. To round off the week, we decided to visit the Mullet Peninsula on the west coast of County Mayo, which was about 15 miles away via the Nephin Beg Range of Hills, standing at around 2,000 feet in height. Approaching the Mullet Causeway, we were delayed by a farmer moving his herd of cows but were soon on to this bleak, flat, relatively speaking, piece of ground much the same as I remembered it back in 1961 with J. Lane, Richard and Anne Symes.

We visited some prehistoric standing stones, beautiful beaches where Jacqueline paddled and Jack sketched, in the south of the island at Blacksod point. Well, all good things have to come to an end, we took Jacqueline back to Knock airport and struck southeast across the centre of Ireland, headed to the Wicklow mountains and a place called Rathdrum, where we booked into a hotel, making friends of the proprietors two children. We toured the area, visiting Bray, a seaside town and Arklow also on the coast and had some idea of climbing Lugnaguilla 3,039 feet, even going as far as to walk up Glenmalar on the approaches but time ran out and we scooted back to Rosslane and home. we had been away for nearly three weeks. Note: this was the last time we visited Ireland.

Around this time, I was warned by Dave Lecoq that if I did not take steps to retrieve my Scott/Norton racer, I might not see it again. So, informing the gentleman in question and using Jack's car we went over to somewhere near Welwyn Garden City to collect it. Upon arrival, we were greeted by his wife and shown a pile of bits! He had taken it to pieces, presumably with the intention of racing it. We returned to Jack's and then took it to Roy Sherwood in Ampthill for restoration. We also attended Paul's graduation at the Philharmonic Hall in Liverpool with Jacqueline.

Our next holiday in July was a real change; we were going 'abroad' using passports – to France no less! We had been invited to spend a week with Rex and June Boyer, old friends from De Havilland's, circa 1960, at his 'Gite', or in English, his country home. We went in our car ALT again, I think this vehicle did more holidays than any other, although we owned the diesels for longer, and we sailed from Southampton to Caen in Normandy. It was quite a long trip, about six hours, not the shortest crossing but put us about 30 miles from Broglie near to where his house was. Caen had been severely knocked about during the war and evidence of this was still visible. Driving on the other side of the road was tricky, although if you remember, I had some experience of it when doing my TA in Germany, circa 1956.

The house in question was a converted barn at St Aubin De Themmey About 1.5 miles from Broglie, with an additional barn, unconverted, as Rex's workshop lying in about an acre of wooded ground. Very nice indeed! (See photos.) The place next door belongs to Gerard a friend of the Boyers; was larger than Rex's and up for sale at £35,000, nothing like this in England, amazing value! We had arrived at festival week and this year's theme was western for which Rex insisted on a wearing a Stetson, cowboy clothes and a six gun and holster! Terrible; his wife just shrugged her shoulders! Mind you, most of the other people looked just as ridiculous. Weather was good, as you would expect and we spent a considerable time drinking/eating at a local bar, owned by one of Rex's friends; typical of him, he seemed to know everybody!

We also visited surrounding towns of Bernay – very mediaeval, Orbeck and Honfleur which had a large yachting centre. Our trip back was reasonably uneventful except on the boat Rex consumed vast quantities of seafood, several helpings, no wonder he was so fat! Just because it was free, well obviously included in the fare. Same approach in the cabin we had; he insisted on clearing all the writing materials, pans, paper, notebooks etc, insisting that we had paid for it in the fare and that this was normal behaviour for travellers – awful behaviour. June didn't think much of it either!

It was around this time that Jack's friend Ted Stone died; he had lived for many years in Africa and had eventually come back to his house in Eaton Bray; he gave great support to our projects and attended several lectures at Cranfield University as I have previously mentioned, we were visiting Jack at Flitton at the time and I went with him over to Eaton Bray to collect some tools etc that Mrs Stone said we could have. Ted's brother's 1938 Triumph Speed Twin was there to be eventually flown out to Australia where he now lived. Well, during our rummaging around we discovered two ball engines (ref photos), which we assumed Ted had made. He was a competent machinist; I don't think they had ever run as no carbon deposits could be seen. We took them back to Jack's house and they remained there until after inquiries, his son requested them, as this was the correct thing to do. It would be interesting to know what happened subsequently.

Three Tors time again, July. I set off from Stanton on the E2 as usual but after getting off the A5 at Woburn Sands, I was obliged to stop as heavy pinking had been taking place for some miles. I reset the ignition timing at the side of the road and continued to Jack's house. Four of us set off the following morning, Steve on his 'Black Pig', Jack 'Dottie', Shed 750cc Royal Enfield constellation, from Steve's house to arrive at Clive Heasman's place in Devises for breakfast; to be joined an hour or so later by John and Derek; on the CSI and 350cc AJS, respectively.

We arrived at Geoff Brown's in the late afternoon after our usual lunch stop between Glastonbury and Taunton on the A361. The following day, Saturday is a pre-run for some and grovelling with their machine for others! (Ref photos.) The usual procedure on Sunday, Brown Willy first, High Willhays second and Dunkery Beacon last. Only one incident of note, when Steve lost his wallet on Dartmoor, to be found later where he had previously parked his bike. Trip back uneventful, at lunch stop again near Glastonbury, I had to reprimand some American tourists for taking pictures of the other bikes, explaining to them that there were thousands of machines like that, but only one with sleeve valves!!

Meanwhile, Fred and I had started on the E3, which was a 500cc water cooled two stroke, sleeve valve engine to be fitted into the same motorcycle as the four stroke E2. Whilst this was our new

250

project, I continued to ride the four stroke around Leicestershire etc eventually clocking up over 5,000 miles!

Next up, holiday wise, was our first visit to the Western Isles in the Outer Hebrides which if you remember, was cancelled in 1994 when the Lane's arrived on their way to Orkney and Shetland Isles! We set off for Sky, as we intended to sail from Uig, as it was a short crossing. I think we did it in a day, as it was late afternoon when we arrived. We had made no previous arrangements about sailing or accommodation etc just went in hope, carrying the tents to be used if required! Ah, the rashness of youth, Age 62 and 61 respectively!

The shipping office was very helpful and suggested we purchase a Rover ticket that would enable us to go to any island for the rest of the season, it was now September, they also contacted the tourist board in Loch Maddy, North Uist and arranged Bed and Breakfast for the night. Upon arrival, we motored across North Uist from Loch Maddy as dusk was falling to a place called Bayhead on the west coast.

There was a hill called Marrival at 750 feet behind us containing some standing stones and a mountain called Eaval at 1,130 feet in front, which we climbed on a later visit, it's the highest in North Uist. The lady running the bed and breakfast turned out to be a district nurse, so had a lot in common with Jill. She gave us some pointers and we spent about three days touring down south through Benbecula and into South Uist but the weather was gloomy, so not all the peaks were visible. We now turned our attention north and headed for the ferry terminal on Berneray, to arrive at Leverburgh on South Harris, in thick fog!

Nothing was visible beyond a few yards from the car and we drove for miles through what appeared to be a giant quarry, huge boulders on either side of the road. Eventually, the fog lifted, and we saw what a beautiful place it really was. We had, as we later discovered, taken the eastbound road along the coast to Rodel, which was a truly rural spot, haymaking in the old-fashioned way. Time was now beginning to run out as we hadn't yet discovered any accommodation for the night, and we wondered if we would have to go to Tarbet.

However, having just gone through a hamlet called Drinishader, standing high up on the shores of East Loch Tarbet we saw a Bed and Breakfast sign. It transpired that it was run by a lady called Chrissy, she had tragically lost her husband a few weeks earlier and was now trying to make ends meet. She was still running the small farm they had built together! We stayed here for two or three days during which time we explored the area and visited a Harris Tweed working mill/loom, very interesting.

The weather was still inclined to be gloomy, (ref photos) but on the third day dawned bright and clear and we decided to climb Clisham at 2,622 feet the highest in the Outer Hebrides. We informed Chrissy who said she would show us the way but would not climb it herself. As usual, we took it head on, scrambling up the rock-strewn east face, to be rewarded with super views from the summit. Time to leave and we did, saying we would return, which we did with Jack a year or two later.

Our way now was to continue north through Tarbet taking the westward road into North Harris towards the Island of Scarp. Whilst we were there we came upon a man, also visiting, whom we had met some years previously, running a book shop in southern Scotland, he had since retired saying there was no future in books with the rise of the Internet, Kindle etc. Interestingly, at the same time we purchased a paper, *Daily Telegraph* I think which contained an article on Scarp (I have a book detailing the history of the island etc *Hebridean Island* by Angus Duncan). Saying that the island was up for sale at £70,000! Marvellous, what a bargain.

We retraced our steps towards Tarbet and branched off northwards towards Loch Seaforth and Bowglass, turning westward arriving at Scaliscro, thence on through Glen Valtos and the Mealisval Mountains To finally arrive at Mealasta; end of the road, where we had lunch and enjoyed stunning views of Scarp from the north. We had to return the same way obviously and turned left at the Gearrald-ha-Aibhe Junction to arrive at the Calanais stone circles with the Isle of Bernara on our left, these apparently being some of the most ancient standing stones in the Hebrides. Further, on we came to the Carloway Broch which is a 'good example of a preserved Broch'. By now it was getting towards evening and the Butt of Louis at Port Ness; where we were able to find another Bed and Breakfast with resident seagull.

Lewis is much flatter than Harris and because of this its two hills Ben Mheadhonach at 556 feet and Muirneag at 806 feet, really stand out and I bet the views from the tops would be very interesting, except that now 2019, there will be many wind turbines visible! Such is so called 'progress'! Instead of cutting down on our energy requirements, we seem to be trying to keep up with the rising population! Crazy! Which leads onto the problem of generating all this extra electricity, especially when and if, electric road vehicles etc, which the government seemed to think is the answer; come into being!

I have always thought the answer is hydrogen, a clean fuel, emitting only water and the way to achieving this is by breaking down seawater, say from the Atlantic Ocean, using the sun's heat from someplace like the Sahara Desert or similar, where it virtually doesn't rain and piping it to wherever it's required, using the existing oil/gas network. This way, we not only reduce or almost eliminating our carbon emissions but can also use our existing vehicles, modified of course for hydrogen. Thus saving the necessity to build completely new machines at vast cost and even more pollution etc. The use of hydrogen is not new, people like Ricardo were running internal combustion engines on it, way back in the 1920s/1930s (Ref. *The Internal Combustion Engine.* Volume 2 – H. Ricardo). There will be huge problems of course and expense and what do you do with the excess oxygen? Is this the fly in the ointment? But all the other alternatives so far suggested, would cost more and achieve less! What price global warming? We stayed here for a couple of days exploring various places like Geiraha on the northeast coast, eventually travelling southwards through Stornoway, capital of Lewis, thence onto the Eye Peninsula to see more standing stones and a chambered Cairn, before returning to Tarbet to sail back to Uig on Sky, thence home.

Our final holiday this year was in October with John, Lyn and Phyllis, to Cornwall. It came about through Lyn's mother, who found an advert in her daily paper for cheap holidays, out of season, on a tourist site at Bosinney near Tintagel; so, she booked for a week's stay. We collected Phyllis in our trusty ALT and set forth from Wellingborough to meet the Lanes at our caravan digs at Bosinney on the North Cornish coast with excellent views out to sea. Whilst there, we visited many sites on Bodmin Moor, did some climbing of Sharp Tor, Carkees Tor, Cheesewring and its stone circles, also revisiting our old friends Rough Tor and Brown Willy. Went to Tintagel and Port Isaac where my Benchini camera ran out of film, 120 millimetres and I had to purchase a cheap plastic 12 off photo camera to finish the holidays with, (ref small photos which were surprisingly good!)

We visited the British cycling museum at the old station at Melford and rounded off the holiday with a long coastal path walk from our digs towards Boscastle which proved a bit much for Phyllis, as she had neither the right shoes nor experience but we carried her back! Good old Phil. To round off the year we had our birthday parties at Iris's home, I was 62, and Jill 61 and the Vauxhall UFO party in December was at the Jolly Coopers Flitton. (ref photos)

And so, say all of us!

1997. In March of this year my attention was drawn to an article on my Scott/Norton in the *Motorcycle Sport & Leisure* magazine by Mick Payne. Investigation revealed that whilst visiting Roy Sherwood in Ampthill he had noticed the aforementioned restoration and subsequent inquiries to Roy and Steve Linsdell, ex-vintage racer himself, confirmed the sequence of events, more of this later! Meanwhile, on the engine front, work was continuing on the E3, 500cc sleeve valve two stroke, a scaled-up version of the E1, 125cc sleeve valve two stroke from earlier years: Fred coming up twice a month usually.

Our first holiday this year was a first also for the fact that it was our first canal holiday. Jill and I organised it via the Ashby Boat Company at Stoke Golding and seven of us set sail in the biggest/ longest narrowboat they had at nearly 70 foot in length – the Bosworth – 68 foot. There was Jack, Lyn and John, The Durrants, Jill and me; none of us had any previous experience except Jack he had many years earlier been as a passenger with some friends; but he had forgotten almost all of it! Most of the 'driving' was done by John, Bob Durrant and me with Jill occasionally having a go, Durrant usually drove too fast, causing other boat people to shout at us; he also seemed to have difficulty getting it through the bridges!

On our first half day, we got as far as Nuneaton, opposite a Chinese takeaway called 'Wing Fat' where we pulled up for the night. I mention this because it was scary; we had all gone to bed, Jill and I were at the stern and, next to the kitchen living area, when I was woken by the movement of the boat. Somebody had stepped aboard and was looking directly at me across the kitchen area. We stared at each other for about 30 seconds or so, neither moving, until eventually he stepped back onto the towpath and disappeared! I eventually went back to sleep, mentioned it in the morning; nobody else saw or heard anything, not even Jill who was alongside me! Typical; they say that Nuneaton is a rough area.

From here we proceeded through the Atherstone locks to Tamworth and its environs; through Fradley junction, where we joined the Trent and Mersey canal, thence on to Great Haywood where we turned the boat around for a return trip, as we were now three and a half days out of our first week. Great views of Cannock Chase were visible to the south of us as we went through Little Haywood spending a night at Manchester, returning up the Ashby canal too early to return the boat, so we went past the boat yard up to Sutton Cheney for the night, returning the boat early Saturday morning.

Meanwhile, trials were continuing with the E2 in preparation for the next Three Tors run in July (ref book), one of the more interesting ideas was the fitting of a Reed valve pack in the induction system between the inlet ports and carburettor. this came about as a result of noticing excessive blowback from the Carb Bellmouth when pulling heavily, say up hills.

The real problem was that the inlet ports were closing too late for the revs in question i.e. racing timings previously! It helped and no more blowback was observed, mid-range performance improved, but top end was limited to about 4,500 rpm! So, I upped the gearing, to get the speed. Around this time, I had a phone call from Mick Payne asking to do another article on the 125cc sleeve valve engine, he had noticed on a previous visit. This resulted in the May issue *Motorcycle Sport & Leisure* magazine, giving a part life history of myself connected to two strokes, i.e. Born near Scott works in Saltaire, apprentice at Trojans, working on motorcycles in the REME etc, also owning several Villiers powered motorcycles, two Scott's and a DKW (Auto Union) car, plus descriptions of the 125 and 500cc sleeve valve engines in preparation, i.e. E1 and E3 at this stage.

He also noticed drawings/photographs of my proposed phased transfer engine; at this time, it was to be a 500cc unit based on two 250cc Yamaha cranks etc; and wanted to do another article on this. I put him off, however by not showing the full working drawings and the technical article I had written in 1978 whilst working at Vauxhall's, because I had ideas of patents which were subsequently rendered invalid when the Scott/Norton MK2 was built in 2008/14. More of this later in the next century!

Around this time, Sarah obtained a teaching post in Turkey, as a result of her training on a Tefl course and was away for four months commencing in June to September, she taught Turkish children English, I believe. We come now to a visit to Malvern and its surrounding hills; we went with Mick and Dorothy, Lyn and John, and Kathleen – Jill's friend from Ealing who used to visit once a year. We went via Calke Abbey for a long weekend, stayed in a hotel in Malvern itself, explored the city, climbed to the highest point of course and went to a Jazz Festival at the Town Hall where I also bought a book on Jazz.

In July of this year, the Scott/Norton, now fully restored, went on display in the Stanford Hall motorcycle museum, as a result of a recommendation by Titch Allen, founder of the vintage motorcycle club, whom I had known for years during my motorcycle racing career, he'd always had a soft spot for specials! At about the same time, I had a letter from a Bob Bunce, (ref letter), who had obviously read the *Motorcycle Sport* article in March; offering to exchange his 1927 Scott super squirrel, for my Scott/Norton.

It was a tempting offer, especially as he wanted to race it, but I turned it down for two reasons.

1. Great nostalgic value – a one off!
2. I had thought that Miles might one day have considered having a go at vintage road racing.

In retrospect, I now see that the machine may well have been outclassed by the 25-year rule etc, so it remained in Stanford Hall museum for about 10 years, until Donnington came along later.

So, we come to, yet another three Tors run with the usual crowd including Miles, only this time with a difference! It was a one Tor run and that was Dunkery Beacon, and this is how I think it happened. For some reason, our usual starting place at Torrington was changed and the start was to be at South Molton service station, where we had an early breakfast. We were then assembled by a VMCC official and told that the room route had been changed for some reason that none of us can remember now, therefore we would set off in single file; whereupon the leading man would stop at each junction, indicate the direction to proceed and remain there until everybody had passed him. Then re-join at the end of the column. Simple! Well, as with all the best laid plans, our group ended up on our own; and when we realised this, it was too late to rectify the situation. So, we had a confab at the side of the road and decided to have a run of our own around the Exmoor area.

It was during this time that we came across a steam rally and village fete; enquiries were made and the net result was that if we would lend our bikes for the display, we could come in for nothing; needless to say we accepted with alacrity and spent considerable time looking at old traction engines etc, one particular example of a two-stroke diesel tractor, proved difficult to start, emitting beautiful smoke rings from its exhaust pipe! every time it fired. Here, I made a classic Collett faux par (mistake) with a capital 'M', I was looking at a BSA Sunbeam, the fat tyred, 500cc incline twin, of about 1950s vintage and explaining to, I think, Miles and Derek, how it wasn't a real Sunbeam (i.e. Marsden) But a poor BSA contrived substitute, when, to my horror, I observed the owner standing right behind me glowering! As a point of interest, this was the first time, after many three tours that Derek hadn't used his faithful 350cc AJS but instead, turned out on a 650cc Triumph Bonneville, circa 1960s. So on to Dunkery Beacon and our only Tor climb this year, where we all gathered on the summit cairn for a photograph in superb summer weather, when I made the statement, 'if this was a sample of global warming, let's have more of it,' Oh dear, it doesn't look so funny in 2019!

In September of this year we returned to the Orkney Isles with Jack, Dave and Sandra Tebbut, none of them had been there before. It started for us in Brighton, we went there for Jo's marriage to Dermos, now held at the Brighton pavilion. Following this, we returned via the Lanes to pick up Jack and return with him to Stanton. There then followed an epic drive for me; we started at 4:44am in the morning and arrived at Scrabster, port of departure at 4:44pm that afternoon, a knackering trip. With only two stops briefly for breakfast and lunch. We went in K743 WEA (weak) only it wasn't! Our first diesel, it already had about 80,000 miles on the clock but was to last us 10 years, taking it to 206,800 in the end, brilliant.

We expected to find the Tebbuts already there as they had less distance to go from the Lake District, but no! So, as the boat was due to sail, we boarded on our own. We had an extremely rough crossing, one of the worst I have ever experienced and that includes the Irish Sea. almost impossible to walk, crockery sliding off tables or smashing, waiters falling all over the place and eventually some cars shifted in the hold. Normally the route from Scrabster goes westward past the Old Man of Hoy, the picturesque rock pinnacle, but due to the conditions, we altered course and took the sheltered route through Scarpa Flow to Stromness. From here, we headed north along the coast road to watch the raging sea, which was so rough that at the Yesneby Cliffs, standing 100-foot-high, the spray was being blown over the top! We eventually arrived at Marwick Bay and a little further on our original digs with Mr and Mrs Clouston; in time for the evening meal at about 6:00pm.

Part way through this came a phone call from the Tebbuts asking for directions, or a map reference. I think Mr Clouston went out and found them as they were fairly close, they of course, had had a reasonable crossing and had seen the old man of Hoy! Typical. I don't remember why they were late, probably organisation, not one of Dave's strong points.

We now started to explore the islands, but in what order I cannot tell so I shall just list the most interesting places we visited. (ref photos.) Stromness of course, a very old shipping port with some excellent museums; Yesneby cliff tops and Scara Brae the prehistoric settlement we visited with the Lanes in 1994; Kitchener's Monument and surrounding cliffs, Marwick Head nearby shows the Kitchener Monument; Birsay Isle again and Earles Palace, now in ruins.

A visit to the black Craig Cliffs at Neblonga which we didn't do with the Lanes in 1994, was extremely interesting with stunning views of Hoy; lots of caves on the nearby beach of Cemetary, which contained old ruins which Jack sketched, all in an area called Dutertown about three miles from Stromness on the west coast.

We also paid a visit again to Rousey Island, not to cycle this time but to look at a series of chambered cairns near Midhow Beach, or rather, Jill, Dave and Sandra walked the mile or so but Jack and I decided; he to sketch and me to read *A Short History of the World* by H.G. Wells. The return boat trip was exciting as the wind had got up, so the crossing was rough, accentuated by the small size of the boat, we did not take the car. Hoy was our next choice, Dave being a mountaineer naturally wanted to climb the highest in Orkney, so Ward Hill it was, without Jack who sketched as usual. This time however, we walked on across the adjacent hills/cliffs to view the Old Man of Hoy from above looking down on its summit – very impressive, we did not lose our binoculars this time!

A visit to Kirkwall and on to Saint Mary's and the South Ronaldsay Peninsula, to visit the Tomb of Eagles once again, I did not go in as it involved lying on your back on a trolley to get into the cairn, not my scene, but they all did it! Somewhere on the island we came across a horizontal water wheel, where the water was directed around one side of the wheel, turning through about 180 degrees before continuing its flow – it didn't seem so effective to my way of thinking as the energy was only obtained from flow speed, nothing from the weight of water in the descending buckets as in a normal overshot wheel! That was also our unanimous decision! Perhaps we had overlooked something.

Holiday over, we returned in calm weather, and Jack was able to see the Old Man of Hoy for the first time taking some photos to sketch from later, these have vanished, so I suppose they have been sold on after his death, to enhance the value of the estate! We returned, like we did with the Lanes in 1994, along the northern coast road round Loch Eriboll in Sutherland, passed Suilvan etc to a Bed and Breakfast in Ullapool for the night, thence home the following day. Our final holiday this year was to the Lake District in the Lanes car.

Lyn had pre-booked two self-catering cottages for two weeks, the first was at Seatoller in Borrowdale and the second at Threlkeld just east of Keswick. This was our first serious attempt to get a grip on the Lake District, all previous visits had been of short duration, so our knowledge was a bit on the skimpy side! However, upon arrival we started on a tour of the surrounding district i.e. west lakes area. One of our first visits was to Cockermouth, first of several, due to its various antique shops and warehouses. The weather was excellent for the entire two weeks, most unusual! So, we decided to climb England's highest, Scafell Pike at 3,178 feet, this would complete the set for me, Wales, Scotland, Ireland, and now England. All four of us set off from the farm tea rooms at Stockley Bridge but after a mile or two, the Lanes pulled out, leaving Jill and me to complete, which we did, returning a few hours later for tea and cakes, well received.

Next up was Angle Tarn, done originally by Jill and her parents, before my time, when her mother panicked at the top, refusing to come down with Jill or her father until a stranger appeared and said, 'come along mother' and got her down without a hitch! The old man wasn't too pleased I understand! However, we approached via Mickleden Valley up Rosset Gill which is the steep rocky slope in question and onto the summit where views of Scafell Pike are visible. We also visited the Langdale Pikes, Wasdale Head, and Eck Dale Valley, over the Hardnott Pass and then climbed up Cleater Moor to the summit which overlooks the Sellafield site. It was now time to move on for our second week at Threlkeld, East of Keswick.

Our digs were in the High Street, right under Blencathra and sharp edge; very impressive, it seemed to loom over us. From here we covered the east side of the lakes. Helvelin area, Kirkstone Pass, Ambleside and up the Sleddale Valley to the GATE Scarth pass. On another occasion, we climbed Cat Bells, which overlooks Derwent water, all reaching the summit without incident. And finally, to round off, we tackled Skiddaw at 3,025 feet where three of us reached the summit, Lyn opted out at half distance, she was entitled to as it was her retirement year at 55! From the NHS.

We rounded off the year for her retirement at a party at Swingate Farm with friends, plus Alan, Holly and Paul. (ref photos.) This was also the year when the Cassini-Huygens mission to Saturn/ Titan took place – more of this later upon arrival in 2004/2005. Also, Jill retired in November at the age of 62½ and this was celebrated with nurse Sandra Merden and two other nurse friends. The whole affair was held in a hall in Barwell, where John Easten played traditional jazz, this time appearing in person with his band, including the woman trombonist who played at our wedding in 1993.

1998. March of this year we decided to return to Michael's at Madogs Wells in Llanfair Caereinion with Jack and Paul. The latter met us there having driven from London in his own car where he was working; one of his first jobs after his gap year and University. He stayed for three or four days during which time we visited Montgomery Castle and the surrounding area, but his real desire was to go back and look at Abersoch, where he had spent much time on holidays as a child.

We did it in a day taking in Bala Lake briefly on route; we looked at Mr and Mrs Williams's cottage at Bryn Celyn Bellaf and the Old Granary, now falling into disrepair and of course the beach where they all played.

After he had gone home, we stayed on for the rest of the week visiting the Clee Hill area near Diddlebury where Jill and I stayed on our honeymoon in 1993 and climbing the volcano, Rodney's Pillar, not far from Welshpool. Finally, a trip to Pistyll

Rhaeadr – Wales's highest waterfall, Jack's first visit. We also were enthralled with the views of comet Hale Bop that appeared this year, the first decent comet I had seen; the latter part of the 20th century was particularly poor for bright comets, even Kahoutek and Halley's Comets were poor in the northern hemisphere!

Next came another canal holiday doing the Stourport ring, which involves starting from somewhere in South Birmingham near Stourbridge, heading southwest through heavily built-up areas, admittedly some lovely old buildings and backwaters; into the Clent Hills into open country to Stourport-on-Severn, where we joined the River Severn as far as Droitwich and then on the canals again to Stoke Prior. Wychbold was our last nights' stop, before returning through the Netherton tunnel at 3,027 yards (one-and three-quarter mile) long, which seemed to go on forever, very impressive.

Riding the E2 regularly, meeting Jack at Church Stow near Weedon on the A5 for an early pint and lunch; we got to know the publican and he would open the bar early for us! This was useful test riding and I was trying out various modifications all the time (ref book), meanwhile work continued on the E3. Iris and I put in protests to try to stop demolition of my old workshop, the blacksmith's barn at Flitton, but to no avail, a new brick building, Mary Anne's studio stands there now. It is reasonably sympathetic and will in time mellow and I suppose, that's what they call 'progress'.

Next up was our first trip to the Forest of Bowland where we had a self-catering cottage with Jack. It's a beautiful area, I believe if ever the Queen retires, this is one of the places she would consider! Whilst there we were lucky, for at the weekend a motorcycle rally took place with the focal point at the car park in the centre of the village near the river, very scenic. (See photos.) We toured the area extensively doing some walking on Crossland Fell whilst Jack sketched. we went to Clitheroe and its Castle, there was a particularly comprehensive sausage shop that we had trouble in getting Jack out of!

We also climbed Pendle Hill at 1,810 feet which is also called the Witch's Hill, it is not the highest but close to, it has a lovely, curved sweep on its eastern flank. We rounded off a lovely week in a shameful manner, in which I was a reluctant criminal accomplice driving the car whilst those two dug up some wild plants! Disgraceful!

Sarah now decided on a gap year and took off with University friends for the Far East, visiting India, Nepal, Malaysia, Thailand and Indonesia – they all seem to have to get it out of their system!

Our annual run this July was the 'scrumpy run', apparently the Three Tors was extinct for some reason known only to the VMCC and those who were getting too long in the tooth; and some who had never been keen on it anyhow (not all took to Richard I think – 'being foreigners' so to speak) it was altered last year, that's why we got lost! This time the same old crowd went, except that Miles was replaced by Nick riding a hired Yamaha or something and Geoff Brown rode a very good looking, pre-war Sunbeam, a proper Sunbeam! The run was essentially around Exmoor, returning along the north coast road visiting Minehead, Lynton and Ilfracombe and we took time off to visit Keith and Maureen Bramwell, as usual. This was to be my final long run of 606 miles on the E2 which behaved itself reasonably except for a seizure at Stow-on-the-Wold on the way home due to slipped ignition timing, again!

The engine was later removed from the frame, which now awaits the E3 and given a final examination before being boxed up until December 2006, when it went to Cranfield for research studies. Now came another big holiday to Raera, similar to the one in 1989; the following attended over a period of three weeks; Miles, Sarah, Jack, Nick, Heather, Lyn, John, Jill and me with Dave and Sandra Tebbut visiting during the second week. The holiday started with visits to Musdale, the Atlantic bridge leading on to Seal Island and the Tigh and Trish pub and of course, Easdale, thence on to the island of Luing, it was of course Miles's first visit.

Another visit was made to Kerrera and as before we walked the perimeter of the island enjoying the views, the weather was superb.

We then took off for Mull, for a day trip, the Lanes declined this one for evasive reasons that we were to discover later. We went in Nick's car with Heather, Jack, Miles, Sara, Jill and myself. The weather had deteriorated by now, but we had a reasonably smooth, early morning crossing to Craigmore and continued south along the coast, stopping for lunch and were rewarded by the appearance of a large sea eagle – very impressive! Finally arriving at Fionnphort for the sea crossing to Iona – too many tourists.

We continued around the west coast heading north past Inch Kenneth where the Mitford girls used to spend their summer holidays in the 1930s and onto the Eos Fors waterfall which was now in full spate (Ref photos). We continue north onto Tobermory, the capital of Mull! And finally, back home to Raera, late evening; the Lanes arriving back somewhere near midnight. The mystery was now solved – they had been to the point of Ardnamurchan, the most westerly point of the British mainland. And why? Well we had been talking about going there on our final third week, when everybody else had gone home. So, to beat us to it, as they were only here for two weeks, they usurped us – no answer to that one!

Glencoe came next, Nick and Heather had a book of walks in the Highlands and this one was listed as grade B, not too difficult but it proved otherwise for the older members. The first half, up to lunchtime was reasonable but the return after lunch became a bit of a struggle. First, the weather deteriorated, second, we had come a bit too far, third, Lows Gully appeared before us. This was a steep, narrow and wet climb around the side of a mountain and Lyn, realising this, decided to take a longer circular route around this mountain/hill or whatever it was. We finally made it back, all a bit knackered, except the youngsters and drove off to try and find Lyn; Eventually discovered miles off down some beaten track; we retired eventually to the Kings House Hotel on the edge of Rannock Moor where they were filming for some whisky manufacturer or other, all very interesting.

Come the end of this week and we all said goodbye to Miles, Sarah, Nick and Heather – all too brief! Our second week started off with a visit to Pulpit Hill above Oban and a walk along the coast to Dunstaffnage Castle for a picnic lunch during which Jack found a £10 note! Extremely useful, we all helped him to turn it into liquor, what else. Lismore Island came next, we walked the length of it and Jill, and I climbed the highest hill. It's only about 200 feet, being long and narrow we did not have to leave the central path in order to see everything. What we had done this week could be described as a local, resting from our previous exertions.

So, we now ventured south along the coast road and onto the Ardfern Peninsula for great views of the Gulf of Corryvreckan, where the famous whirlpool is situated, more of this later, when we returned to Jura with the Tebbutt's. Continuing south along the Sound of Jura in the mist, we arrived at Keillmore, a lunch stop where Jack did some sketching, before turning east towards Inverinan on the shores of Lough Awe, with a visit to the Ardanaiseig Hotel, where five teas cost £10, no cakes, more of this later! Later this week, the Tebbutt's arrived on their way home from the Highlands and a good time at the 'Trish Trash' was had by all.

So, to our final week, just the three of us. We had already decided to visit the Ardnamurchan, in the wake of the Lane's, as it turned out and so headed north to the Ardgour/Corran crossing, across Loch Linnhe, with spectacular views of Ben Nevis. We headed south taking the A884 to Lochaline, the ferry to Mull and Kilundine Castle, again, where Jack did some sketching and onto Drimnin Where the road ends. Return by the same route is the only car option, so we arrived back at Strontion, turned on the

A861 west and arrived at Pollach near Salen on Loch Sunart, where we found a bed and breakfast for the night and very good it was. The following day, it was on to Ardslignish for a view of Ben Hiant at 1,712 feet (we didn't climb this one) although it looks to be the highest on the Ardnamurchan.

We carried on the B8007, arriving at Kilchoan Castle and the ferry to Tobermory on Mull, where again Jack sketched, finally reaching Achnaha on the outskirts of Sanna Bay, where we found our second nights bed and breakfast with a Mr and Mrs McPhail, superb! It was hereabouts that we came across a strange rock formation near the beach, that we named 'The Wellington Bomber' due to its fuselage geodetic construction, the likes of which this rock seemed to resemble! Jack photographed and sketched it, but I have no idea what happened to them after his death, we would like to have had them (particularly the sketch) to hang on the lounge wall.

We finally arrived at the point of Ardnamurchan and its lighthouse (ref photo). It was here that Jack and I saw a huge seal come 'sailing' by which merely gave us a look of disdain, so much for humans. We spent a couple of days or so here exploring the area during which time, I bought a book at the local post office called *Night Falls on Ardnamurchan* by Alisdair Maclean, Detailing the lives of the people who lived here in the area all those years ago! Inevitably, we moved on, following the northern coastline to Fascadale Bay, where we did a lot of walking, the weather the whole time was excellent, even on the one occasion we walked back across the Moors to Sanna, an eight-mile round trip! We were fit then, only 63 and 64. The views from here were outstanding, to the north – Rum, Eigg and Muck well visible and to the west – Moidart Hills, sea and mountains everywhere.

On to Kilmorry a remote little village, then returning via the B8007 and A861 to Ardmolich on the River Moidart and Glenuig, for our third nights bed and breakfast in a 'bunkhouse', a large common room with several beds and a separate room to eat in, rather like a youth hostel, except we didn't have to bring or cook our own food.

In case you were wondering, we solved the sleeping arrangements by Jill going to bed first whilst we were out of the room and then facing the wall whilst Jack and I got into our beds, fortunately there were no other visitors, I can't remember what happened in the morning! We had to repeat this procedure several times on some of the remote holidays. Whilst here we made a further visit to the coast at Smirisary to see some deserted coastal croft's, remnants of the old clearance days and on the roadside came across an old crapper' – toilet to you! It stood in a field over a stream so that the effluent was carried away! Brilliant, we didn't try it. We were now well out of Ardnamurchan and into the Moidart region and on the road to Mallaig where we turned south to come home via Fort William etc as our time had now run out, to end what was a wonderful time, with people you love!

Our last trip this year was in November, when I noticed an advert in the *Astronomy Now* magazine, very similar to Michael's in mid-Wales, offering bed and breakfast plus telescopic views of the cosmos, over in Norfolk, dark skies etc. So, Jill and I set off, to North Norfolk, I don't know exactly where but it was within range of Holt and Southwold. We met a Doctor Batty who seemed to be in charge, plus wife and had a good time viewing all kinds of celestial phenomena, in a dark rural situation, also visiting Sid King's Motorcycle Museum in Holt, where he had moved to after living in Shefford for many years. It was a short week which ended with a visit for nostalgia's sake to Southwold where we booked a night's bed and breakfast and had the most enormous kippers for breakfast – similar to the ones at Lochranza on Arran.

1999. This year started with our usual visit to the Lanes for new year and involved a long walk on New Year's Day around Faversham and its environs with Rod and Lyn Tindall and their relations. Our first holiday was to the Eden Valley with Jack, who proposed it, also visiting Edna and Norman at Kirkoswald, their first home here. We went via the Peak District and through Yorkshire, Richmond etc, and visited some stone circles at Little Salkeld in February – another early start!

This was followed in May by another canal trip with the usual seven of us; this time it was organised by Bob Durrant and he chose the Grand Union Canal starting at Leighton Buzzard. We went north through Linslade where the Great Train Robbery took place some years earlier, through the Blisworth tunnel to Astcote where we saw a hot air balloon come close to landing in the canal, passed Bugbrooke (Aunti Louie and Tom) And through the locks just south of Weedon where the canal crosses the A5. On most of the photos you will notice the glorious sunsets! We had excellent weather, not so the following year!

We continued through Norton and Braunston, the canal centre of England, as far as Knapton Before we turned back for home visiting the Canal Museum at Stoke Bruerne and unbelievably seeing the Goodyear airship near Bugbrooke at almost the same place as the hot air balloon on our way up!

The arrival of June bought another trip to the Western Isles, this time with Jack, Dave and Sandra Tebbut, it was their first visit and our second. The three of us went via Uig on Skye, to Tarbert on Harris, as before, to arrive at a pre booked self-catering cottage at Ardhasig; a couple of days before Dave and Sandra Tebbut were due to arrive. We used this spare time to revisit Chrissy McKay at Drinishader on South Harris and introduce her to Jack, she in turn introducing us to a newborn lamb! Also revisiting Horgalost and Luskentyre beaches on the West coast (ref photos).

The next day, when they still haven't arrived, we set off for Mealista, this time intending to climb Grosival, at 1,615 feet, an interesting peak nearby and left Jack sketching. The weather was good, and the climb strenuous, but enjoyable and we reached the summit about half an hour later, all's well so far! From the summit we could see now that Dave and Sandra had arrived, we had left a note, so decided to take the quick way down and we nearly did! We set off down the north facing slope which gradually got steeper and steeper and slippier and slippier, until even we realised that all was not well. With the loose scree we did have a bit of a job on our hands and knees to get back up! Stupid, my fault, not the first time I have taken risks, but not to involve another – least of all, darling Jill; shades of Galtymore! The boy never learns!

The Tebbuts were with us for the first week and during this time we visited Rodel on South Harris where more Skulduggery was performed – shades of Forrest of Boland previously. This time it was an old, deserted cottage near Rodel that they found, to discover like in Ireland, that the previous occupants had left, leaving almost everything behind. Well our merry crew of vagrants decided to remove a few items such as crockery, cutlery etc and use me again as the getaway driver – such is life! Other trips included a trip to see the island of Scarp, a climb up Ceartaval at 1,807 feet and a revisit to the Carloway Broch and a black house on the beach at Great Bernera.

At the end of the first week we now split from the Tebbuts to go to our second digs in a seafront cottage in Amhuinnsuidhe, on the road northwest towards Scarp; whilst they went south to visit Benbecula at South Uist etc. As soon as we arrived, we were adopted! A large friendly collie dog; funnily enough similar to Jack's Judy, arrived and Jack spent the rest of the week feeding it every day (ref photos). Amhuinnsuidhe is an interesting place, having a Castle, waterfall and Tirga Mor at 2,206 feet highest point in this vicinity.

Let's start in that order, first the Castle; we paid it a visit for afternoon tea and cakes – Jack's idea of course and met his Lord Puffum! Yes, a real Lord, or so we were told by the staff. He seemed a nice chap (not guy) very friendly, slightly vague, his age I suspect; nevertheless, it all went off very well. Next, the waterfall, very impressive, recent rain had improved its flow and it was going like a good 'un! Finally, Tirga Mor at 2,206 feet. Which Jill and I climbed, surprise, surprise! With its wonderful views of land and sea in all directions and bought back some interesting stones of various shapes, sizes and colours for Jack.

From here we toured north to the Butt of Lewis, over to Tolsta on the east coast to see the waterfall at Tolsta Head, Where Jack and Jill seemed to be plagued by midge's – all that fuss! Reviewed Ben Muirneag and Ben Mheadhonach again, the northern most hills on Lewis. On to the Eye Peninsula to the Tiumpan Head lighthouse, which we now understand to be a 'dogs home'. We returned to Tarbert for our return trip home but took time out to visit Scalpay Island to view a concrete ship, apparently used during the war to transport munitions etc, it's the only one remaining and Jack did some sketches of it.

This year's Three Tors was slightly different, it took place in August on the eleventh when we went down to Geoff Brown's for the last total solar eclipse of the sun in the 20th century! Viewed from Bodmin Moor. The full details of this event have been previously mentioned in my review of the Three Tors when working at Vauxhalls. To finish our holidays for this year, Jack, Jill and myself visited eastern Scotland because Jack reckoned that we spent so much time on the west side that we had totally neglected the east – Good point!

Our first week was spent at Tarland on the east side of the Cairngorm mountains in a bungalow on the edge of the village. Whilst we were there the car exhaust pipe fractured but fortunately the landladies husband ran the local garage and he effected the repair. We spent a day in Aberdeen mainly watching the shipping whilst Jack sketched. On one of the trips we passed a farm that had various machine tools standing outside some large barns. It proved too much of a magnet to pass, so I went to investigate and ended up purchasing a German made boring head with a number five Morse taper which fitted my horizontal milling machine, for £50.

Our second week was spent at Cuminestown further northeast, about 20 miles from Fraserburgh and was again a single storey building, not a bungalow but a very long old-fashioned cottage in the High Street; of this much smaller village. We had a trip along the north coast road and visited the hamlet of Crovie which Jack was much taken with, his idea of isolation, it would have been a bit tricky in bad weather as the cottages were virtually on the beach with high cliffs right behind them, no shelter whatsoever! From here we went to Gardenstown and this is the only photograph (photo ref?) that survived from this holiday, and it was taken by Jack, I have no idea what happened to the rest and to ours too! A visit to Fraserburgh Castle and lighthouse museum proved very interesting, it's amazing how many varieties of lens were used in lighthouses etc!

Whilst here, we went south to the Loch of Strathbeg near Rattray Head, to be amazed at the number of sea birds, there must have been thousands, if not tens of thousands, the noise was 'awesome' – a horrible modern expression! A large proportion of them were geese, gathering, I imagine, for a migration south as it was now late September; Jack was in his element, photographing and sketching to his heart's delight – what a shame all this has been lost! I hope that somewhere someone is enjoying the sketches that were probably sold by his executors after his death. *Note:* contact with Leonard in 2020 has revealed that he has now moved to Greenfield and has got all of Jack's diaries and sketches and has promised access, should I require it. Great!

On another trip, we went to Peterhead and then to Cruden Bay to visit Staines Castle an ancient monument perched right on the cliffs edge, thence on to Ellon and across the hills back home, thus ending a glorious two weeks of good weather! For Scotland!

Finally, we were invited by Linda and Rod Evans to visit their holiday home in Rustington, which lies halfway between Worthing and Bognor Regis on the Sussex coast. We went down with John and Lyn; Jill and I slept on the lounge floor and inhaled some of the fag ash that lay there, I remember the smell, they were both heavy smokers! Later, Geoff Gunter arrived in his MG Sports Car and we all went home after about two or three days thus saving us from possible lung cancer! Finally, I officially retired in November to receive my old age pension after 65 years of toil, by a grateful government! (My comments on the pension scandal are unprintable) almost! Perhaps later.

Marriage to Jill - March 1993. Hinckley registry office.

Off on our wedding night - Stoney Stanton.

David Berry - September 1993. Our old school.

Fred and I at 'Stone Eng' May 1994 with E2 500 cc 4-stroke sleeve valve.

E2 500cc 4-stroke sleeve valve - 1994.

Lynn, John and Jill cycling around Rousay in the Orkney Islands - May 1994.

Holm - of - Skaw. Most northerly point on Unst, Shetland Isles - 1994.

On our way back from Orkney and Shetland at Applecross
with John, Lyn, Rob and Jill.

Precelly Mountains, South Wales. Jill, Rob, Lyn and John at Pentre Ifan.

Rob and Jill on the summit of Goat Fell, Arran. 2840 feet - October 1994.

Alan's 'Pap' on Jura - May 1995. L to R John, Jill, Jacqueline, Lyn, Alan.

Lyn and John at cottage in Ardara, Donegal - October 1995.

280

Jill on the summit of Errigal.

'Quiet man café' Cong, Connemara. Lyn, Jill and John - 1995.

Miles racing at Milton Keynes. Jill, Iris, Sarah (Miles' girlfriend),
Jack and Miles - April 1995.

On the 3 tors at Devises. L to R Jack, Clive, Steve,
Derek, John and Rob - July 1996.

The 3 tors - 1996. John, Derek, Shed, Jack and Steve.

Ball engine found at Ted Stone's house - 1996.

Second ball engine in foreground. Origin unknown - 1996.

Lynn, Jill, Rob, Phyllis on Bodmin Moor - October 1996.

Jill; say no more.

A tough climb - Jill on Galty Mountains, Ireland - 1996.

The summit of Galtymore looking West - May 1996.

'Coming home' – Jill and Jacqueline in County Mayo - 1996.

Nevin conquered - Jill and Jacqueline - 1996.

Having breakfast at Lettermaghera. Jacqueline, Jill and Rob - May 1996.

Jack's friend in Lettermaghera. Mrs Boyle's house in County Mayo.

June, Rex, Jill at Broglie, France - 1996.

Picnic at Mealista, Harris - September 1996.
Grionaval at 1615 feet in the background.

Swingate farm - August 1997. Alan, Paul and Holly.

Bob and Margaret Durrant, Jill, Rob and Jack on the Bosworth - May 1997.

Great Haywood junction.

'Captain Jill' in charge on the canals - May 1997.

1 'tor' - 1997 L to R John, Steve, Shed, Jack, Derek, Miles,
Geoff and Ann in Goodleigh, Devon - 1997.

Dunkery beacon at 1686 feet. 'If this is global warming,
let's have some more of it!'

'Cat bells' in the Lake District - October 1997.

'Madogs Wells', Wales - 1998. Me, Jill, Josephine, Dave, Sandra, John Easten.

'Raera' - 1998. Seil Island near Easdale. John, Jill, Miles, Lyn and Sarah.

Glencoe picnic - September 1998. Nick, Heather,
Lyn, Jill, John, Sarah and Miles.

Jack, Jill and dog - it waited every day for Jack to feed it.
Cottage at Amhuisuide, Isle of Harris - September 1999.

Chapter Thirteen

21st Century and – More Holidays –
2000 – 2020

The holiday season started early this century with a visit to the black mountain area of South Wales, with Jack where we visited the ruins of Castle Carren, toured around and climbed Pen-Y-Fan At 2,879 feet, the highest in South Wales, finally overlapping into March when we revisited Michael and Ann at Madogs Wells again, for a few days, before setting out in April on the canals again. It's all go, you know! Makes you wonder how Fred and I ever got enough time to make engines! Progress had continued on the E3 with a trial fitting of empty castings into the frame to assess the engine plates and water connections and radiator; for this engine was, unlike the four stroke E2, water cooled, using a Yamaha radiator and a canal boat central heating 12 Volt electric water pump, thermostatically controlled via the thermostat sensor in the cylinder head. This also meant that of course there would be no Three Tors this year, for me.

P.S: Nobody went anyway as Geoff Brown was emigrating to Australia.

The canals by now the responsibility had returned to us again so we chose to start from Napton and head down the Oxford canal, which is very winding, few locks and therefore takes a lot of time and miles to make any headway! A small incident here, almost caused embarrassment when I spied a pile of scrap metal which included some twisted/broken propeller shafts which would be useful in the workshop. So, I asked the woman in the office if it was OK to remove them; on the affirmative, placing them in the car boot. Upon our return a week later, she informed me that inquiries had been made about the loss, and suggested we remove ourselves as quickly as possible – point taken! However, I have got ahead of myself; we set sail up the Napton locks, the last we were to see until Banbury. All went well and we got as far as Aynho just north of Oxford and returned the same way, taking our time. It was the shortest canal trip we made, I think we were getting a little stale because it was some years before we went again. We, meaning Jill and I.

This time it was to be Jura in May/June and it was the Tebbuts first visit and our third. The three of us, Jack, Jill and myself set off for Tarbert for the sailing but arrived late and had to have a bed and breakfast at Kennacraig nearby on the Mull of Kintyre, with a Mrs Dixon who turned out to be an interesting character, like Jill, who was very interested in all things including gardening. They corresponded for some time afterwards even sending each other various items of mutual interest! We arrived the following day at Craighouse and drove north to our pre booked digs at Kiels nearby. Later on, with Dave and Sandra, we drove up the only road along the eastern side of the island to Ardfernal and Lagg where Jack did some sketching whilst we all explored the area, noticing that the deer seemed to be unafraid of humans, even to the extent of roaming around the village! (Ref photo.)

On another occasion, we visited the Invar coast in the west to satisfy Dave's whim to see the raised beaches and caves thereby – all very interesting. Now, of course we come to the main event, certainly in Dave's eyes – the climbing of the Paps! To coincide with the annual Jura Pap fell run, where all four Paps are covered in the minimum time, to and from Craighouse. Well, this is how it all went wrong for him and Sandra; they decided to go to Isla for the day, don't ask me why? I can only think they got the day wrong as we did also; and also, unfortunately the boat to Isla sails from the Feolin Ferry right on the south of the island and not Craighouse where all the competitors were arriving from the mainland!! Meanwhile, we, unaware of all this drama decided to do our own things, i.e. Jack to sketch and Jill and I to climb the small pap and one other.

So off we go, me with my army motorcycle greatcoat, ridiculous – idiot! we tackled the small pap first with no trouble except that we kept seeing people with little on, rushing past us in the opposite direction. And so, it came to pass, that we eventually reached the top of one of the major Paps, to find a group of observers, who had been watching our slow progress with amusement that anybody could or would climb in a bloody great coat like that! Needless to say, the Tebbuts were not amused when they returned.

The next combined effort was a visit to the Corryvreckan Whirlpool situated between the north coast of Jura and Scarba. It is caused apparently by a pillar of rock on the sea floor, that under certain conditions of flow the water is given a spiralling motion, thus causing the Whirlpool at the surface; it usually occurs around the equinox's, so I am told. It was a rough trip as the last mile or so had to be done on foot, through tangled undergrowth and plenty of midge's, Needless to say the Whirlpool was not operating that day! Following this, a trip down to the south of the island by Jack, Jill and myself to view an impressive waterfall near Loch Bhaile Mhargaidh, led onto a walk near the coast where we came upon the remains of a crashed aircraft! (Ref photos) At a farm called Cnocbreac.

Subsequent inquiries from the farmer, revealed that it had crashed into the Paps during, or just after the war, during a training flight and it had not been economically viable to remove the remains. Later investigations by myself suggests that it was a Fairy Fulmar, Naval training aircraft with a box like fuselage, twin seater, low wing, powered by a Rolls Royce 1145 HP Merlin engine, powering a three bladed propeller. (ref photos again) This was, of course from our point of view, the highlight of the holiday, at least for Jack and I and again he sketched it – but these have been lost, at least to us for the moment.

After this we returned home via Isla, visiting a woollen mill at Rutha Ban; Bowmore the capital, and several more interesting places, before sailing back from Port Ellen to Tarbert for another overnight stay with Mrs Dixon at Kennacraig.

In September, we took off again, this time to visit southern Scotland at Sanquar in Dumfries/ Galloway with Jack. This was another pre booked self-catering house. set in the hills overlooking the town of Sanquar on the A76. We were to be here for the first week and then go on to sail from Ardrossan to Arran for the second week; But those bloody mice turned up again!

All went well to start with, we visited several places locally, eventually deciding to go to the lead mines and beam engine at Warlockhead. Whilst looking around, Jack fell, and subsequent inquiries revealed he had broken his ankle! The people at the lead mine museum provided a wheelchair and we got him to the car, WEAK diesel, and drove to Dumfries hospital, where he was admitted immediately; operated on and placed in a bed on the top floor with good views of the surrounding countryside, and a very interesting companion; a man who had spent his whole life working on the roads in Scotland! This happened in the middle of the first week and we visited him every morning of that week.

What to do next? Crisis point: we had booked the place in Arran for three people and it seemed a shame not to use it as such, so inspiration, one of us, I don't remember who, suggested Charlotte and she drove up in her car to stay with us for the remainder of the week, leaving her car here and coming in our vehicle across to Arran; Her first visit and our second (see 1994). Our cottage at Thunderguy was a menagerie; chickens, ducks, geese, donkeys etc, you name it and it was there! Amazing. The gentleman that ran it was a Mr McMaster who lived on the premises and had some sort of leg impediment that demanded he walk with a stick.

The setting was brilliant, almost on the beach, looking west with the Mull of Kintyre in the background, idyllic! The menagerie extended indoors, in the form of insect life of all shapes and sizes; Charlotte found a few spiders in her bed initially but we soon got used to checking these things out first; there were compensations; one of which was a lovely peat fire every evening, lit by Mr McMaster and fresh eggs, milk etc. There was a beautiful lake on the slopes of Bean Bhreac at 2,323 feet which was accessible from the back of our digs; an evening walk we used to take on occasions.

Whilst all this was going on, we took time off to write to Jack in Dumfries hospital, hoping to cheer him up a bit – I hope it helped; we were later to take him back here in 2001 but enough of this later (ref also the letter, among the photos). We toured the islands by car visiting many places of interest, it's very beautiful and easy to get to, as I said before, it's a miniature Scotland. During these visits we went to Lochranza distillery where Charlotte bought a bottle of Arran whisky – the price I notice was more than you would pay in a supermarket!

Two trips stand out in memory, both dependent on the weather. First, a walk up Glen Catacol; it started well but as time advanced rain appeared, getting progressively heavier until it was sheeting down. We scrambled up several difficult gulleys, until we reached Loch Tannar and called it a day! We were completely soaked; my Wellington boots were full up with water and had to be poured out! The total distance for the round trip there and back according to Charlotte's Pedometer was eight miles! Why do we do it? We were young then, only in our sixties! Second, a walk to a stone circle and historical remains at Newton on the Cock of Arran with the sound of Bute in the background; the weather was superb this time and many wild deer were seen also; they tend to congregate in the north as it is less densely populated and mountainous. Our return was uneventful, Charlotte picked up her car and we went to Dumfries hospital to pick up Jack and take him home to Flitton.

We had telephoned in advance of course, suggesting that as we were returning south, we should take him to save an ambulance and after a cursory examination they agreed. We completely lowered the near side front seat back so that as he sat in the rear, he could have his leg out straight in front of him and Jill sat in the offside rear seat alongside him; good job it was a large car with all the luggage in the boot. When we arrived back in Flitton he was immediately taken in hand by the lady opposite who insisted on doing everything for him – what a popular fellow he was, so we spent one night there and returned home after what I can only described as an eventful two weeks!

Visit to Tom and Sylvia's for a week to look after the place as they had gone to see Tom's sister in Switzerland. We were visited by Lyn, John and Holly who came with one of Derek's grandchildren.

To round off the year – we had a bad winter with lots of snow, and Mick had just acquired Ellie, a Collie Bitch, one of his quieter dogs and we still owned the large Orchard field behind Mick's paddock but now decided to sell it to him. This came about because Jill had never wanted to own this land, she only bought it to stop the brothers from selling it off to goodness knows who! The annual maintenance i.e. cutting the grass etc was irksome. So, after a few years, we agreed to sell it to Mick whose land it adjoined.

2001. This was the year that foot and mouth made an unwelcome return, in February whilst Jill and I were on a three-day visit to Michael's at Madogs Wells and another bed and breakfast in mid-Wales, I cannot remember where. This farmer announced it to us and also went on to explain why he preferred solid fuel boilers and central heating, to electric, gas, oil alternatives. His reason: that there was always a large central hot core, that retained the heat, a good point as at this time we were on a Coke mix and logs ourselves at home. We paid another visit to Woodbridge with John, Lyn and the Durrant's in May to explore the local area including a trip to Pin Mill on the River Orwell, which included some interesting old boat/ship relics from bygone ages – very interesting indeed.

Our next trip was to return to Dalton Hall near Richmond in Yorkshire with Jack who had now recovered from his broken ankle, and Dave Tebbut who had just broken his! Well almost, it was his Achilles tendon on his right leg that he had snapped! And it came about thus: after last year's visit to Jura, 2000, he had decided to return alone, as Sandra was working and could not get the time off; and walk the entire roadless west coast from south to north, a distance of about 40 miles, over very rough country indeed, camping out when and where necessary! He had, of course done this on many occasions previously. Well, as I understand it, he had completed two thirds of the distance and was north of Loch Tarbert when he was looking at some bird or other, he stumbled, fell and felt a savage pain and that was that! Just shows what a knife edge we all live on. At this point he would be about six or seven miles from the eastern coast road and some help.

So, what did he do? Well, the sensible thing, of course, he erected his tent, cooked a meal, slept overnight, cut himself a stick, repacked some necessities, leaving the surplus food behind, and headed east. I don't know exactly how long it took to reach the coast road but the going must have been hard over such rough terrain. The first car he saw coming was driven by some old ladies who refused to give him a lift! Incredible, but I don't know what he looked like! However, eventually a vehicle arrived and took him to Craighouse and a Doctor who immediately sent for an aircraft to transfer him to Glasgow hospital. This meant that Sandra had to take a plane to Glasgow, then public transport to Tarbert on the Mull of Kintyre, where he had left his car at Kennacraig and drive it back to Glasgow, pick up Dave and drive him home, leaving the rest of the stuff on Jura for any fortunate visitors who happened to pass that way!

To return to the narrative, we drove to the Yorkshire Dales, stopped briefly at Richmond for lunch arriving at Dalton Hall in the afternoon, it was our second visit. In the following days, we toured the area for a while visiting Tan Hill and on to Swaledale, to Ivelet Bridge which is very similar to the Atlantic Bridge on Seil Island; (Ref photos;) usually ending up at Greta Bridge Hotel for tea and cakes! Whilst here we also visited Norman and Edna at their new home in Croglin in the Eden Valley.

Meanwhile, work was continuing on the E3 and the engine was completed and fitted into the frame for its first test run in April and ridden around the test route for the first time in May, to complete a total of 428 miles by the end of the year! (Ref engine books.)

This June, Jill and I decided to visit the North York Moors by ourselves and set off northwards without making any prior arrangements, except putting a tent and sleeping bags in the boot, such is the confidence of youth! At the age of 65 and 66, respectively. Eventually, we somehow arrived at Grosmont and found a Bed and Breakfast for the week. It contained some interesting people including a motorcyclist from Denmark, who had ridden over to watch the making of *Heartbeat*, which of course had our own Bill Maynard, who lived in Sapcote, in it. All of this was new to Jill and me, we had no idea about soaps, but we did go over to Goathland nearby to watch some filming; but didn't see Maynard or anybody else of note. We did however take a trip on the railway that runs from Grosmont to Pickering in the south and a lolly good trip it was, return of course!

One incident that deserves a mention, was the derailing of an engine at Grosmont late one afternoon and the subsequent night's work required to get it back on the tracks. This went on all night and as our bedroom faced that way, we could see the arc lights blazing away most of the night. Apparently, we were told, this had to be done quickly, so that the management and the press at the Pickering end would not know! Further visits to Rosedale Abbey, Bransdale and Roseberry Topping at 1,046 feet which is a very impressive peak, visible for miles, were made. No, we did not climb this one. The weather was good all week and it was nice to be on our own for a change.

We returned to celebrate Miles and Pat's wedding on 28 July, at Wilstead Church, followed by the reception at the Cranfield Halls where an Italian style meal was provided for all and sundry. The weather was exceptionally hot, so not much dancing, more drinking to cool off. Jacqueline stayed overnight with Jill and I, at Jack's house, the only time she has ever visited. We went later this month to meet Richards parents, Arnold and Valerie in the Peak District; in preparation for Jacqueline's wedding that was due on 6 October in Manchester.

This next holiday to Scotland had three objectives: -

1. To visit Wigtown – Scotland's Hay-on-Wye – books!
2. To climb Merrick at 2,766 feet – southern Scotland's highest peak.
3. To take Jack to Arran to visit the menagerie at Thunderguy.

So, we booked two places in advance – a cottage/bungalow at Stairhaven and Mr McMasters place in Arran. We roared north and got to Stairhaven in one day; a nice little modern bungalow, on the hills above Luce Bay, with great sea views to the north, west and south of us. On almost our first full day we headed for Wigtown on the A747 coast road via Auchenmaig Bay and Mochrum Loch, looking at a ruined Priory and Chapel on the way. The book shops at Wigtown are all mainly on the High Street and although comprehensive, there are nowhere near as many as Hay, but nevertheless!

We all came away with something – mission completed! With our usual singlemindedness, we now headed for Merrick which lies in Galloway Forest Park; a rather gloomy start off. It lies at the centre of the Range of the Awful Hand, let me explain. If you spread your hand out on a flat surface, Merrick will be where your second finger knuckle is and the digits of your hand represent the spurs of land radiating away, sloping downwards, naturally as Merrick is the highest point; and there are five of them ranging from southwest to northwest, very impressive!

We started the climb from the Bruce Memorial and Glen Trool Lodge visitor centre, leaving Jack to do some sketching etc and for the first mile or so up Buchan Burn we were enclosed by pine trees; after Culsharg Bothy, we broke clear and had good views all the way up to Benyellarg at 2,358 feet and so on up the ascending Nieve of the Spit to the summit, a distance of about three and a half miles. The high rounded bulk of Merrrick dominates the hills around Loch Trool and this 'branched finger' is the highest summit in the Galloway Hills. From here, on a clear day it is possible to see the English Lake District, the Arran peaks, and the Mourne Mountains of Ulster, and we saw at least the latter two! We returned the way we came with a group of 'Octogenarians' who had incredibly done the whole seven-mile round trip, I hope we shall be as fit when, or if, we get to our eighties!

Sadly, it was not to be for Jill, but I wonder now (2020) if I could do it? Possibly given enough time! Further, the detailed information given in the text of the climb etc, comes from the books by Cameron McNeish & Richard Else called *Wilderness Walks* which accompanied the BBC TV series; of which I have both parts, one and two.

A visit now to the Rhinns of Galloway, that odd Peninsula of land that exists to the west of Stranraer, a place I was familiar with after trips to Northern Ireland, of course. We toured around, visiting Portpatrick etc., where Jack was much taken with a Reliant three-wheeler van, with girder Forks and an Austin seven engine and rear wheels (ref photos). He would have very much liked to sketch it, but the owner returned and was in a hurry to drive off – pity!

From the Mull of Galloway in the extreme south we could clearly see Slieve Donard etc in the Mourne Mountains, Ireland. Now it was time to move on to Androssan via Girvan on the coast road A77 viewing Ailsa Craig at 1,105 feet, standing about 11 miles offshore from Girvan. On our arrival at Brodick we took the northern route to Thunderguy and introduced Jack to Mr McMaster and all of the various animals, some of which he sketched during his time here.

We covered most of the island in one way or another whilst we were here, viewing seals at Merkland Point, a trip through Glen Chalmadale where deer and eagles were visible. We had a return to Glen Catecol and re-walked some of it – in fine weather this time, thinking of Charlotte! We also toured the southern half of Arran to visit Lamlash on the east coast where in a field below The Ross at 1,010 feet, we saw a group of horses enjoying themselves, running around their field!

It was on such a visit, returning from Blackwater foot when we decided to call in on a cafe near the Auchagallon stone circles, late afternoon for tea and cakes, to receive one of the greatest shocks of our lives – 9/11. It happened thus; we entered, sat down at the table and started to examine the menu, when a side door burst open and a man came out saying 'come in here and look at the TV, I can't make it out, is it a horror movie or what?' or words to that effect. While the rest is history! This was late in the week, Thursday I think, two days later we took the boat from Lochranza to Claonaig on the Mull of Kintyre to take the long way home, around the Mull which Jack hadn't seen, and on through Bute, Durness, Glasgow, etc and back home. It was whilst we were on the boat from Lochranza that it was stopped in mid-stream, engines off, just drifting in the current, whilst the captain said a prayer for the victims of the tragedy, and we all observed a minute's silence!

Jaqueline's wedding comes next on 6 October and what a bizarre affair it was; it was essentially in two parts, first the civil wedding for legal reasons as the 'Church of Christ' was apparently not recognised in this country! And then the second ceremony for the Church of Christ. Well, the first part went off OK, signing the register etc in the presence of witnesses but during the ceremony of the second part the officiating vicar or whatever name he goes under in this Cult, started to make some disparaging remarks regarding Richard, that the audience took exception to, especially Arnold, Richards father, who found the whole affair disagreeable, as I did myself, and expressed ourselves by growling/fidgeting etc, murmurs of discontent! This stopped it and the rest of the ceremony passed off peacefully. A lot of the families and friends found the sight of Cult members praying in groups, holding hands, weeping and wailing etc, very distasteful and were glad when it was all over. The venue was a large hotel on the outskirts of Manchester near the airport and an evening buffet and dance were provided. All went well until about 10:30 pm when the fire alarms went off, and we all trooped out to stand outside the front entrance for about half an hour; not very funny for those people who had already gone to bed! I don't think we ever found out what the cause was.

Another visit to Dalton Hall, this time with the Lanes in their car, now October, they had not been here before and found it quite to their taste, bed, breakfast and evening meal and all that. We did the local area as usual, visiting Richmond, along Swaledale to places like Thwaite, Muker and the Falls at Keld over the Buttertubs pass to Hawes for some of their Yorkshire cheese! We also went as far west as the Howgills; this is an area north of Sedbergh that contains more mountains per square mile than any other in the UK and was visited by Dave Tebbut on one of his forays when he was caught out in the open by a thunderstorm and took cover under an overhanging rock, very dodgy! The big waterfall of Cautney Spout is visible on one of the photographs; and then back to Greta Bridge Hotel for tea and cakes; not all done in one day I hastened to add, but over a period of a week.

Finally – the faithful three, Jack, Jill and myself returned to Madogs Wells in November just as Michael had finished two of his projects.

1. He had landscaped some of his newly acquired lands to the north side of his house, putting in a footbridge etc for the convenience of the residents; Jack spent quite a bit of time sitting in the sun and sketching.
2. He had set a yellow globe representing the sun on one of his gateposts and measured out to some scale or other, the

distances of the planets of the solar system, so that people could follow the indicators way up the hill behind his property, finally to Pluto which was about a mile away! Quite a task, Jill and I walked it; on the same scale, Alpha Centuri the nearest star would be in Australia! Fearsome. Most of our time was spent, as usual touring around to visit and/or revisit places like Pistyll Rhaeadr; whilst here we had a visit from the Tebbuts and John and Josephine Easten whom we took to see the Pistyll Rhaeadr waterfall– information supplied by Sandra Tebbit 2020, and a very nice cafe in Carno. The year was rounded off by a meeting with John and Rosemary Collett when they brought their new narrowboat up to a Marina at Kilworth where we saw them with Phyllis, I think she had come for Christmas. (Ref photos.)

2002. We started early this year in February with Jack, the Lanes, Jill and I; so much so that when we arrived in Kirk Yetholm at our pre-booked self-catering farm cottage, it was snowing hard and continued to do so on and off throughout the rest of the week. However, in spite of this we toured around successfully, in our car – a bit cramped, five up, but never mind. This was a cottage actually on a farm, so a fresh supply of milk and eggs etc was available; the views to the front of the Cheviot Hills were wonderful, particularly when snow topped.

Many trips were made locally to Kelso bookshops; Wooler where great views of Dunmore Hill at 1,842 feet and the Cheviot itself at 2,652 feet were seen, we would, that is Jill and I, like to climb the Cheviot; highest in the area, but the rough going, long distances and icy conditions, sensibly restrained us. We went east over to the coast and viewed Bamborough Castle and into Beal where the crossing to Holy Island can be made twice a day/night at low tide; we also went north onto the Moorfoot Hills where views of the Pentland Hills near Edinburgh were obtained.

About this time, I joined the Mars Society and was allotted an acre of land on Mars!! (Ref my Land Deed!) Still it's good for a laugh – somewhere to go if this global warming thing takes off! Still, I received the quarterly bulletins for a few years until I stopped paying the annual subscriptions and got a couple of books from them. Meanwhile, on a more serious note Jean, Charlotte and Jill had taken allotments up, next to the church Cemetery in Knock Verges, at the same time as housing development were taking place in Station Road where the old 'Cuture' factory and scrapyard had stood. The man involved was Blore, owner of the new Triumph motorcycle factory on Hinckley's ring road and he also wanted to build on the allotment ground – greed! They attended many council meetings, fought him and the council, tooth and nail, to eventually win, helped by a national drive to encourage allotments that was going on at this time. Good on them; the allotments now in 2020 are really thriving, every plots taken, and an Allotment Society has been formed.

Next up was a visit to Kinnerton in mid-Wales where John Lanes' pal from his RAF days, national service, lived. We went in separate cars to their beautiful cottage and impressive garden, to find them both in good health. During this time, we climbed Bach Hill at 1,982 feet, the local high spot, visited Devils Bridge and took a return trip to Aberystwyth on the miniature steam railway, also visiting the Observatory at Knighton. On the latter part of the week we went south to Hay on Wye, Hay Bluffs to visit Llanthony Abbey again, Michaelchurch Escley, where Jill and I had previously had bed and breakfasts, returning through Abbey Dore, the Golden Valley to Breadwardine where Reverend Kilvert is buried! Jack had visited this area many years before to see this. Contact with Alan and Margaret from Kittle Green on the Gower, South Wales, involved us in a visit to Black Mountain to climb Gareg Lwyd at 2,008 feet; they having met us halfway, approximately – all very nice, a good day out.

In between times, I was with Fred of course and continued to ride the E3, to get it ready and reliable for the next Three Tors. (Note: this was not a Three Tors in any way, as it took place in Norfolk and was the last time that all the old crew (i.e. Jack, Steve, Shed,

John, Derek, Peter Wheeler and myself were to ride together). Whilst Fred and myself started work on a variable timing arrangements driven by epicyclic gears situated in the timing cover, that would by rod and lever operation, enable the rider to change the phase relationship, hence the valve timing whilst in motion. More of this next year! Unfortunately, during this time Richard's father Arnold died of a sudden heart attack in June; surprising as he appeared quite fit, having done a lot of walking in the hills and retired early. So, Jill and I visited Glossop for two days for his funeral etc.

Back to the Outer Hebrides; six of us, the Tebbuts, Jack, Charlotte, Jill and myself, but not all in one go! No that's too simple! Dave Tebbut started first, he set off for Harris and Lewis about a week before Jack, Jill, Sandra and I; and arranged to see us in Lock Boisdale South Uist the following week. So, we for set off up the A1 to Oban for the sailing. We only got as far as Catterick before we ran into an enormous traffic jam, but as we were creeping along in the slow lane, a small, what looked like a farm track appeared and we decided to take it. Progress got worse as we bogged down in a muddy puddle/lake resulting in, "everybody out and push" which didn't go down too well! However, after a circuitous route along back roads we rejoined the A1 ahead of the jam.

Oban was reached in the late afternoon, where a very reasonable bed and breakfast was obtained – only £14.00 per night per room, Jill offered to double up with Sandra to reduce costs further but was politely turned down! Later, instead of our usual fish and chips on the seafront we went to a posh seafood restaurant and paid double? And a seagull did its business on the bonnet! We sailed the following morning, and all went well until we reached the Minch and open sea, where a long swell pitching motion took place, causing Jill to be sick, fortunately into the *Daily Telegraph*! She said it wouldn't have happened if Sandra hadn't asked her to go on deck and see the view!

When we arrived at Lochboisdale in the late evening, Dave was waiting for us and after a local meal we retired to our pre booked digs (ref photos). Several interesting places were available within walking distance of our digs, one was Fiona McDonald's grave 'Our Lady of the Isles' who incidentally was born here on South Uist. Secondly, there was a hen harrier who regularly flew low across the fields to the south of us and had probably a ground nest fairly close, we did not find this but Dave spent some time searching! Thirdly, we all spent considerable time in the late evenings, walking along the machair (raised beaches) peering into the long grass to see corncrakes which are also ground nesting birds; long gone from the British mainland due to intensive farming etc we often heard their 'Creake' but only occasionally viewed them. On another occasion, we visited a chambered tomb, close to Mingearaid, where Dave and I decided to climb Seabhal at 724 feet and to quote Dave; 'as we came over a ridge, we disturbed an eagle and saw another in the air, also a red grouse'. Jack was pleased to hear a skylark singing.

After a day or so here we received a phone call to say that Charlotte, our sixth member, would be arriving at Benbecula airport from Glasgow the following day. Once she had joined us, we visited further afield to places like Ormiclate beach where Jill and Charlotte spent time paddling in the sea, for what? (Ref photo!) At Loch Sgioport they found some wild ponies to feed and a further visit to Loch Ainort resulted in climbing Trimeabhal at 643 feet, whilst Jack sketched, as usual.

Somewhere around the middle of the week the Tebbuts announced that they were going to climb Ben More at 2,015 feet, the highest in South Uist, but with Charlotte and Jack we thought it best to abstain, so we went South to the Ludag Ferry, the crossing point to Eriskay hence Barra; passing in the process, a very old turf covered cottage. A real gem for Jack to sketch, (ref photos) I bet it's not there now – too good to last! Finally, a trip to North Uist with Charlotte to visit the ruins of Cairinis which I think was the remains of the clearances many centuries ago and view out to sea the Haskeir Islands in the Sound of Monarch and Dun Fort at Griminis.

At the start of the second week, Charlotte had to leave us, and we all motored up to the airport to watch her plane bank eastwards heading back to Glasgow. We now decided to get even with the Tebbuts; so, saying nothing we took off with Jack to Cairinis in North Uist with the idea of climbing Eaval at 1,128 feet, the highest in North Uist. So, leaving Jack sketching, we set off for what was a very wet walk, dodging various ponds and lakes that dot the countryside around here as can be seen on the map. This went on until, after about a mile and a half, we arrived at the foot of the mountain and started to ascend, through a narrow valley sloping upwards that took us to the summit, very beautiful. To rub it in we also went southwest to Creagastrom in Benbecula and climbed Ruabhal at 403 feet, the highest on Benbecula! Our final climb together was Benscrien at 1,251 feet on Eriskay, the island reached by a bridge now, that lies between South Uist and Barra, where the film *Whisky Galore* was made in the great days of British Ealing comedies! Plus, a visit to the Arm Politician pub!

29 July: The final Three Tors. As explained, the initial problems were that Shed, had previously broken his ankle, was driving somebody's Porsche car and Steve was riding somebody's BSA sidecar outfit with Peter Wheeler in the chair! Otherwise! We set off from Stoney Stanton after Jill had furnished us with a full breakfast; everybody stayed two nights; and headed east through Uppingham to Deeping St Nicholas on the A1175 towards the Wash coast near Long Sutton where we inspected an old Victorian pumping station, before turning back for home. All went well, E3 running nicely, until we reached Tur Langton, when John's Norton developed a puncture; fortunately, very close to a pub which was open late afternoon Saturday (ref photos). The total mileage for the trip was 235 miles in very hot weather, not as demanding as a normal Three Tors of 700 odd miles but rewarding nevertheless, for me as the bike went perfectly, no oiled plugs or anything!

A quick visit down to London on the River Thames for Mary's wedding to Clint took place; the thing that stood out for me here was a woman harpist who played beautifully, I do like to hear a harp, it is a very restful sound to my ears! That's one of the great moments in the Marx brothers' films, which are mostly rubbish, when Harpo plays his harp! Sorry about that I got carried away!

Following all this, a trip to the Lanes, with Jack and a visit to Brooklands with Derek also, for a VMCC event where we met Simon Grigson and Sandy who live locally, when in this country. It almost ended in disaster for me; with the usual Collett stupidity, I decided to climb up the racetrack banking to look over the other side, there was nothing to be seen of course, except trees, however when I set off down again, I went too quickly, found myself running faster and faster, until I overbalanced, just managing at the last moment to put my right arm up to save smashing my skull! Apart from tearing my trousers, jacket and cutting my knee, bruising my ribs etc, I got off lightly – what a twit! At least I can say 'I've crashed at Brooklands'. To round off the week, we visited Tom and Sylvia for eating and drinking.

Mull, wettest of the Scottish island was now to be given a thorough visit of two weeks; previous visits had been only transient and much of the interior was unknown. So, Jack, Jill and I set sail from Oban to Craigmore once again, it was now September. We had pre-booked digs at Frachadil near Calgary on the northwest coast, they were run by a middle-aged lady who lived in her cottage next to the converted barns that served as our dwellings – very nice too!

We toured around virtually the whole of the island road system, returning 'home' at the end of each day, so I'll just record some of the more interesting points. We took a trip to Tobermory, the capital, very picturesque with all the differently coloured houses and from here we took a boat ride to Kilchoan on the Ardnamurchan peninsula for old times' sake!

At Dervaig, where we used to go shopping for provisions and a cup of tea/coffee, I found that the proprietor had originally worked in the British film industry and had on display more or less all of the studio film books that I have spent years collecting i.e. Warner Brothers, MGM, RKO and Paramount etc. Except one, he didn't have a copy of 20th Century Fox, very rare, I've never seen another one, mine was donated to me by Pete Clarke, second in charge of the drawing office at Hunting's, a keen movie buff! Unfortunately, the gentleman was losing it, I suppose the dreaded Alzheimer's disease, so he was not able to give me as much information as I would like – what a shame. Easfors Waterfall was revisited on the west coast, followed by a trip across to Ulva and Geometra where we had fine views of Ben More at 3,147 feet, Mull's only Monro, which we subsequently climbed, to its full height from sea level, a long hard, not too difficult slog! Photos of Jill on the summit show her to look a bit disgruntled, not her usual cheery self! I think we were a bit knackered. Still the views were terrific and the weather good, all two weeks.

We also took a boat trip to Fingal's Cave on Staffa for the first time for Jack and me; Jill had previously visited in 1989 in a trip from Raera with Lyn, Sarah, Tom, Sylvia etc. Remember? Following this up, we visited Iona again our first visit, where Jack and I spent some time in the book shops, whilst Jill went on a walk around the island by herself; beautiful place but spoilt by excessive tourism because of the religious connotations. On our way back we visited Erraid, a village that was part of the setting for R.L. Stephenson's novel Kidnapped also visited the Ardmeanach Peninsula and Bearraich at 1,404 feet, for some spectacular sea-cliff views and the 'Drovers Inn' at Grass Point and Carsaig.

An interesting visit to the Macquarie Mausoleum near Salen revealed that Macquarie was the man who explored the northern territories of Australia; born on Mull, they done him proud! We also covered most of the castles on the east coast and went to the Little Theatre, 36 seats, near the Drimnacroish Hotel, to watch a comedy play whose name I forget. The return home was uneventful.

2003: Throughout the latter half of 2002, Fred and I had been working on an epycyclic gear arrangement to give variable port timings as the engine was running and this was fitted early January of 2003. The whole contraption was controlled by a lever, hand operated by the rider, which was fitted to the end of the sun wheel countershaft. To see the details of this, refer to Jack's beautiful illustration of the E3 or of course the main drawing – general arrangement, GA.

It gives a total timing range of 40 degrees from before bottom dead centre (BBDC) to after bottom dead centre (ABDC) such that all reasonable timings can be explored. At the same time, believe it or not we started work on the E4 – where did we find the time? Several test runs were made between January and March which indicated that improved performance was achievable between load and RPM by varying the lever position, such that with a servo motor fitted controlled by sensors monitoring these two factions, much improvement would be possible, things were looking good!

Alas, whilst coming through Thurlaston on 11 March, there was a shattering bang and lock up, fortunately at low speed and that was that! I called Jill from a nearby phone box and she brought the car and towed me home, we didn't have a trailer at this time.

Subsequent examination showed that the planet wheels had broken, we suspected brittleness, by being overheated and, caused by water quenching, leading to further damage to the piston, sleeve and the rest of the gear train. (Ref book of E3.) This arrangement was not tried again, by common consent; as the amount of work involved, mainly by Fred, was deemed undesirable! But variable timing continued in other ways both on E4, E5 and E6 engines later in 2004/6.

First holiday this year was to the far North of Scotland at Kirtomy, a hamlet near Bettyhill in Sutherland. We three, set off, having pre-booked a self-catering cottage right on the edge, almost, of a cliff overlooking the sea; and drove as far as possible before reaching Inverness and the Black Isle area, where we successfully located a Bed and Breakfast for the night. For our evening meal we had the local fish and chips and we all agreed, they were the best we had ever tasted – just thought I'd mention it! We set off the following morning and got about 20 miles up the A9 before Jack said 'here, I haven't got my teeth in', so we returned, much embarrassment and set off again and about the same distance out he said, 'here, I've left my overcoat behind!' This time we carried on agreeing to phone them up and collect it on the way home!

Denis Murphy – Jack's old schoolmate, boyhood friend and fellow fighter pilot – fell into hysterics when it was all told to him – he always liked a good laugh! We arrived at last and set about exploring the area; local trips to Strathy Point, with superb views east and west along the north coast towards John O'Groats and Cape Wrath, respectively, were rewarding, followed by visits to Ben Loyal and Ben Hope again.

A visit to the Altmaharra Hotel for a midday pint proved interesting; we got talking to a fellow Englishman and friend, who it turned out, knew Stoney Stanton. His father lived close to Ashby Parva and had about 40 motorcycles. I immediately knew him, I visited several times and used to see him regularly on his stall at Stanford Hall every July's VMCC meeting! What a small world. It turned out he was climbing hereabouts and has just returned from Ben Kilbreck at 3,116 feet and was going to climb Ben More Assynt at 3,243 feet the following day, he had also 'done' Ben Hope at 3,012 feet, thus completing the set of the most northerly Munros.

Whilst exploring the north coast close to 'home' we came across a sheep that seemed to be stuck on a ledge about 20 feet down from the top of the cliffs, with no apparent way of getting off. Whether or not it had fallen we were unable to ascertain but there did not seem to be any pathway. We thought about reporting it but as usual did nothing, except vowing to return the next day and it was still there, then report it – to whom, we weren't quite sure, as it was a remote area! However, it was gone on our return, goodness knows how it achieved it – magic I suppose or alien intervention!

Our furthest excursion eastwards was to view and possibly climb Morven at 2,291 feet near the coast at Dunbeath, and on the way down to Straith of Kildonam, we came across a 'goldmine'. The notice outside read to the effect that if you paid a sum of around £10, I think, they would supply the equipment and site for a day's 'panning for gold' in the River Helmsdale. 'Keep whatever you find,' we did not indulge!

However, we continued to Dunbeath and along a small track towards Morven, ending at the Braemore River, with still a four-mile track over rough moorland, across Maiden Pap at 1,573 feet, a total of nearly eight miles round trip. By now, time was moving on into the afternoon; so, we abandoned the idea, I don't think Jack would have appreciated hanging about all that time either. One of my fondest memories of this holiday was just sitting in the rear conservatory, (ref photos) with Jack and Jill, watching all the wildlife, during the evening until dusk – no television for us. On one occasion, we saw an eagle fending off two crows that were trying to relieve him of his evening meal!

We travelled from here mainly on the coast road for our second week at Polbain near Achiltibuie, another pre-booked self-catering bungalow set down a steep flight of steps toward the beach. We had some trouble with this as the landlord had previously enquired whether an 80-year-old person, i.e. Jack would be safe on these steps, but Jill convinced him, after some discussion and assurance from Jack at all that would be well! After his broken ankle, we kept a good eye on him when traversing dodgy ground, etc. Polbain proved to be an excellent choice, with a general store and Post Office etc., just opposite our digs and the landing stage for boats to the Summer Isles just adjacent; so, we took a trip over there, about a mile out to sea to Tanera the largest and wandered around finding Fraser Darling's house.

Who is Fraser Darling I hear you ask? He was the lighthouse keeper on the Farne Islands off the coast of Northumberland and on 7 September 1838, he and his daughter Grace, rode through a storm to the wreck of the *Forfarshire* and saved nine lives, for which they were awarded a medal for bravery. We did some local climbing but nothing of any consequences until we visited Stac Polly which proved too much of a magnet for Jill and me. We set off from the car park at Inver Folly on the shores of Loch Lurgain, thus climbing the full 1,989 feet from sea level as was done last year on Mulls Ben More. except this time Jill did not reach the summit; I only just managed it, as the final approach was a rock climb, hand over hand very similar to Tryfan- (ref photo from summit).

Two other visits came to mind, a trip to Ullapool and to the gardens at Leckmelm, more of interest to Jack and Jill, I think I went and had coffee! Then to the impressive waterfalls of Measach where we came upon a Scott motorcycle rally – what luck. We spent some time talking to the competitors, who were doing a tour of Scotland – good luck to them!

Two notes of depression followed:

1. Rex Boyer died of prostate cancer; we attended his funeral at Lower Stondon with Ted Snook; he was an old friend from my De Havilland days, the first of them to die!
2. Steve was knocked off his motorcycle at Snetterton, badly breaking his ankle, so this removed any further Three Tors outings this year; there had been talk of another Norfolk tour.

In June, we took off for the North York Moors again with Jack and Charlotte, to stay at a farmhouse, converted, self-catering barn in Rosedale, which had some very interesting old lead mines, within an evening's walking distance and a barn roof full of cats! (Ref photos.) Interestingly, also were barns full of cattle, which looked a bit tatty, standing in water etc that Charlotte took umbrage too and was going to complain to the farmer? Her hearts in the right place, bless her.

We visited an interesting old church in the Bransdale Valley, then moved on to Rievaulx Abbey, for a packed lunch in the gardens above, eventually taking the B1257 to Hawnby with its quaint hotel and then on northwards for some good views of Roseberry Topping near Ingleby Greenhow, where Jack managed to acquire yet another cream bun! (Ref photos.) Later in the week we went to Whitby; lost Charlotte for a while, I think she went walkabout; to view some remarkably interesting old sailing ships in the Harbour, finally sampling some of their delightful fish and chips! 'Food of the gods' as Jack called them! Of course, we had to go on the railway again, so a visit to Pickering and a look at the Castle, followed by a journey to Grosmont, lunch by the river and return. Finally, a visit southwards to Nunnington Hall with its beautiful gardens and a classic afternoon tea and cakes, supplied by the lovely waitress – Jill!

Our final holiday this year in September was almost a disaster from the start, the players in this drama were as follows: Jack, Jill and myself, innocent bystanders, Tom and Sylvia going also to their son Matthew's wedding, and Lyn and John who provided most of the action. It happened thus: we picked up Jack as usual and brought him to Stoney Stanton the Lanes followed intending to stay the night and leave with us the following morning, however, John was suffering from rectal bleeding and lost a lot of blood in the night so they decided to return home for examination and treatment etc. they left us Tom and Sylvia's wedding clothes that they were taking up to Scotland for them as they were travelling up by train.

We went on our 'merry' way with Jack and the clothes arriving safely at Glen Sligachen on Skye, with instructions to deliver the clothes to the Ardanaiseig Hotel, Loch Awe, where the wedding was to take place; a place incidentally we had been to before and found it very expensive! Nevertheless, it was done. And funnily enough on our way back we pulled into a garage for fuel, only to discover the bride and groom were at the adjacent pump, so we explained all, sadly having to rule out the Lanes completely – or so we thought; it gets better! Meanwhile, Tom and Sylvia having arrived at Tyndrum Station had got themselves across to the Hotel and the wedding took place.

We three returned to Sky to receive a phone call from the Lanes, saying they were coming up after all and would pick up Tom and Sylvia from Tyndrum Upper Station and bring them later in the evening, to join us! It appeared that they had got little support from the medics at the hospital, so John said, as they had paid for the holiday, he was determined to have it. Good for him – takes guts! So, we had two photos of all seven of us to commemorate the achievements, which you may have noticed, has taken over a page of this document to relate; without having said a word about the holiday yet, I told you it was a saga!

Now, what did we do during the next two weeks. Our house was situated at the head of Loch Sligachen with the Black Cuillins behind us and the Red Cuillins in front and to the southern side – very spectacular. We took the following trips, but not necessarily in the following order. Heading south, we visited the beautiful beaches at Elgol and Tarskavaig, Where I had my annual paddle in the sea to wash my feet – so they said.

The otter sanctuary at Kylertrea, close to the ferry from Glenelg on the mainland, which we had previously used, was interesting for those interested in giant rats! And of course, a visit to Glen Brittle, right in the heart of the Black Cuillins, where we could see the footpath of five miles, back to our digs at Sligachan, instead of a car journey of 10 miles! Nobody availed themselves of it. Whilst all this was going on, John was having the 'bleeds' mainly at night in a separate building called 'The Bothey' and it's a credit to them that we were more or less unaware of the situation! Except that at times John looked very pale.

A further trip north to view and walk in the Quirrang was made, with views of the Old Man of Storr, a rock pinnacle of slender beauty, and visits on the north coast to Stein, and return to our old friend Uig, scene of many sailings to the Western Isles. One day, we all took a sailing across to Raasay, the island made famous by Flora McDonald when she looked after Bonnie Prince Charlie, in a cave on the southside which we did not visit, as our intention was to go and see Callum's Road in the north. So, we set off and walked (no cars here) about four miles towards Arnish where he had lived. He was a remarkable man, an individualist! Callum McLeod lived at the northern tip of Raasay, miles from the nearest road. Over the course of 50 years he watched as neglect and decline took their toll on his small community.

So, he decided to do something about it. He built a road. One spring morning, he set off with a pick and shovel and wheelbarrow and alone began with his bare hands a quixotic venture that would dominate the last 20 years of his life and his road would become a powerful and beautiful symbol of one man's 'defiance against the erosion of his native culture'. This is a quote from the book *Calum's Road* by Roger Hutchinson, written in 2006, after our time here; I don't know quite how we became aware of this road, discussions with John and Lyn seemed to indicate that there was a plaque near the jetty at Raasay telling us about it. I thought we had some precognition of it from either Jack or Tom – who knows, four out of the seven of us are gone! 2020. Anyway, we reached it and walked about two or three hundred yards along it as a tribute to the man, I wish we could have done it all, but time and tide were running out!

One morning at breakfast, when it had been decided the night before that we would again go south through Elgol to Loch Coruisk, Tom had his 10-minute egg! The 10-minute egg was the time of hard boiling, that Tom insisted on having his egg done! Rock hard! That riled Sylvia so much as to exclaim, 'why did I marry an idiot'. He declared his intention to go north to Portree by bus, to get some money from his bank, he used to do this if you remember in Oban from Raera. So, we revised our plans such that Jack and I went to Portree with him and continued north to explore the north coast; whilst he said he would prefer to get a bus to Elgol, which he duly did. When we arrived back at the digs, they returned in John's car having walked around the shore of Lough Coruisk, well into the heart of the Black Cuillin.

Up to now, we hadn't done any real climbing, and this had to be rectified; so, Jill and I settled on the highest in the red Cuillins, Glamaig at 2,512 feet a Corbett no less! It was also handy – within walking distance of our digs at Sligachan and when we reached the top, after a hard, constantly uphill slog, we could wave to the others at the digs down below! The day we left to come home was eventful for bad weather, the first we had really experienced all two weeks, so much so that I had to empty the car footwell of about six inches of water from the previous night's rain, it improved as we went south.

We had one of our first visits from John Symes in November, I had re-made contact with him, after a lapse of several decades and went to pick him up from Rugby Station. We spent most of the few days he stayed with us visiting Twycross Zoo, he was always very keen on animals and smoking his Tom Thumb small cigars, in our garden room, Discussing old times etc – very nice. I enjoyed these sessions and Jill stayed out of the smoke! He came several times afterward each year until his death at the age of 75 in 2007.

2004. Engine wise, things got even more complicated if that's possible, as we were now working on rebuilding the broken E3, manufacturing the E4, which we started in 2003 and in July, I started on the E5. Crazy! However, first things first, we had to make a new sleeve, piston etc; fortunately, the original timing gears were still available and the E3 was now rebuilt with these. The idea now was to try and replicate what Roger Cramp was doing with his desmodromic, uniflow, scavenged, two stroke, i.e. an early closing exhaust, just after bottom dead centre, to trap the fresh charge in the cylinder, instead of the usual loss down the exhaust pipe. By a strange coincidence, a year or two before this, Titch Allen had turned up one day, with Roger Cramp in tow after visiting Stanford Hall. And it was then, that I first became aware of his desmodromic, 125cc, two stroke engine based on a Honda bottom half; my own work at this time of course, was to achieve uniflow with sleeve valves.

But, and here is the strange bit, I was able to show him drawings of a 700cc opposed piston, two stroke, using the E3. 500cc bottom half, with a 200cc homemade upper half, that controlled the opening and closing of the exhaust ports! (Ref drawings.)

But, this was not a normal opposed piston two stroke, as say the Junkers Jumo or the Rootes TS3 engine; as the upper half was inclined at 20 degrees from the vertical lower half, to give a compact combustion chamber, with substantial squish area, exactly the same as Rogers engine, but of course larger and without the desmodromic head. This has cemented our friendship over the years since we first met, of course when we vintage raced together all those years ago; and it's been a privilege to know such a resourceful and clever man.

However, the E3 had to wait as the E4 was being installed in the frame ready for testing; subsequently, it was not to appear until the first half of 2005. Later, and we started work also on the E5 in this July! Briefly, the E4 was Schnurle parted (hence not uniflow) and scavenged, same as all non-deflector piston, two strokes to date; but the sleeve now operated 180 degrees out of phase compared with the uniflow E3. this was done for two reasons: -

1. it provided very rapid port openings, as piston and sleeve were moving in opposite directions, i.e. not in phase.
2. It counterbalanced the piston motion to about 70%.

It first ran on 25 July – in the presence of Jack, John, Derek, Snook, Mike Lane, Jean and Jill; and the following day was ridden to Stanford Hall VMCC founders day rally, with substantial backup! (Ref Photos.) It aroused a great deal of interest, many questions, an offer to buy and a cheer when it started – first kick to go home. After this, it went to the BSA owners meetings with Brian Herbert each Monday evening at Walton, and many other outings, also back to Stanford Hall Scott Owners rally in September with Geoff Brown and his wife Ann who were over here from Australia for their son Richards wedding; to record a total of 603 miles before a strip down for variable timing modifications which took place in 2006.

Also, at this time a surprise development occurred regarding the 125cc sleeve valve two stroke E1. It came via Fred who was now working for 'Marle Powertrain' in Northampton, a subsidiary of the main works in Germany. It appeared that his employers had a link with Cranfield University, and they wanted to do some work with sleeve valves! And had approached Marle to see if they could help, which of course they couldn't. But somebody got wind of this and apparently said to one of the bosses, 'there is a bloke who works here who makes sleeve valve engines', so Fred was asked all about it, the result being that a meeting was set up in Northampton to discuss all the details. There were about seven of us; Fred's boss, Hugh Blaxill, Derek Lowe an independent engineer, bought in by Fred's boss for support, a Doctor Vaughan – head of mechanical engineering at Cranfield, two students, Fred and myself.

It all went very well, and we agreed to resurrect the E1. The 125cc sleeve valve engine (at this time remember, the 500cc four stroke was in a motorcycle with the Es 3 and 4, doing road runs) and loan it to them for whatever evaluation they required! Fred kept in touch, visiting Cranfield occasionally as the work demanded. Somewhere near the end of the year the con rod broke and Steve (Flitwick motorcycles) repaired it at Marle's cost. Testing then continued into 2007. Three technical reports eventually arrived and make interesting reading; a comparison with the Rotax 125cc, two stroke shows that in spite of a 30-year difference in ages, our engine stands up very well in such parameters as: Power output, torque, fuel consumption, etc.

Meanwhile, we managed to fit in a holiday or three; down on previous years for obvious reasons. First off was a return to the Cheviots with Jack and Dave Tebbut, who now – his Achilles tendon was repaired – did quite a lot of climbing (see his Independent Record of this holiday sent by Sandra in 2020) we were here one week. We stayed at a cottage in Roxborough, not far from Kirk Yetholm, where we previously stayed with the Lanes. Visiting the start/finish of the Pennine way, Kelso for books, Wooler, Melrose, where Dave went off to climb the Eildon Hills; Hethpool where he climbed Great Hetha with Jill and myself.

We all took local walks along the River Teviot, sometimes to Roxborough Mill which Jack sketched, of course. On one occasion, we found a dead pheasant that Dave photographed, and Jack sketched back at the digs. We did not eat it, as nobody seemed to want to cook it – what a shame. I'm sure it would have been alright, it smelt okay!

Next came Port Henderson, Wester Ross, Scotland with Jack and the Tebbuts, including Sandra who had to get time off work. We had booked a bungalow on a Peninsula between Loch Gairloch and Loch Torridon with spectacular views in every direction, weather permitting, which on this holiday, was not often, much to Dave's displeasure as he had come expressively to climb high mountains which abound in this area. We avoided the worst of it by concentrating our efforts to the coastal regions, where it always seemed better. Eventually, after several abortive forays where they got soaked, Sandra rebelled, and he joined us on several outings; not before however, he had seen the transit of Venus through his prepared optical filter. Let me explain: every 122 years with eight years between the next two transits, i.e. in 2012 there will be another! A transit is when Venus goes between us on the sun, so it appears as a small black dot on the Sun's surface, it's an eclipse actually and lasts an average of an hour or so.

Starting with local visits, we were much taken by Red point, which lay at the west end of our cul-de-sac road, it had marvellous views of northern Skye and Raasay, we could almost see Callum's Road! Turning east, we shopped in Gairloch and during a visit to the gardens at Iverewe, we ran into Norman and Edna who were on a visit from Scourie, where they were holidaying in their caravan. With the weather being grotty at times, we spent quite a lot of time on the Melvaig Peninsula, surrounded by the sea, visiting places like Cove and Inverasdale etc and quite a lot of tea rooms and bookshops. We also went South along Loch Maree to Kinlochewe and down Glen Torridon, very beautiful, to places like Alligin Shuas and lower Diabaig with extensive views of Loch Torridon and Applecross.

Of course, we had to do Applecross, especially as Jack hadn't been there, and as there is only one circular road around the coast, we took it, had lunch in Applecross itself. Same as with the Lanes previously and finally got as far south as the old Strome Ferry that Richard Symes and myself used on our way up to Durness in 1957! It is now bypassed by bridge and road, and finally to return home.

Sarah's wedding came next in August, which was held near Crouch End – London which is where they were living at this time – not far from Karl Marx's grave! All went well – they dressed me up in a monkey suit, with top hat etc, from Moss Bros; Gareth, being a Roman Catholic meant that the service was a bit too religious for my liking but there you are! Can't be helped. Afterwards, we all drove about 20 miles to near Burnham Beeches for the reception, Jack and I shared a room overnight; for the wedding breakfast, Rosemary Collett smoked and supplied cigars, whilst her husband John dismantled the lounge clock, as apparently it wasn't working; much to the consternation of the staff, who were happy to see it back in one piece – still not working!

The Lanes did not get much sleep, as the adjacent boiler room plumbing made so much noise; they complained, hoping for a rebate but were given a free night at a later time – I don't know whether or not they ever took advantage of it! Photographs of the wedding are available upon request! This next holiday in September came in two parts, the reverse of 2000 when Jack broke his ankle; by starting in Arran and finishing in Sanquhar! also there were new players, i.e. Jean, Clive and Mary Heasman, plus of course the usual three!

We stayed, for a change at Blackwaterfoot, Drumadoon Bay, on the southwest coast, in a rural, off the beaten track, self-catering cottage of the old type. We did all of the usual places we had done before, so I'll not describe them again. (ref photos). One interesting point, however, was different. We used to visit the large hotel in Blackwaterfoot which was just down the track from our digs, and around the walls in the foyer and lounge etc were many paintings of excellent quality. Of course, Jack made for them, explaining to us ignoramuses, what they were all about etc. After several visits, he announced that he thought several might be originals and should perhaps not be so conspicuous but fought shy of saying anything to the management. Trusting to the fact that most people wouldn't know anyway, including probably the staff – ignorance is bliss!

For our second week, we returned to Sanquhar and our previous digs outside the village with Jack and Jean, who were travelling in our car, whilst the Heasman's returned home to Devises. We toured around as usual, showing Jean the sights and taking Jack back to the Wanlockhead lead mines, where he broke his ankle, Just for old times' sake! Later in the week, Jean's son Jamie and his wife Orla came to visit from Edinburgh for a few days. We did several long walks with them, mainly out the back of the property, away from the Sanquhar direction, over the open moors, which was like a high plateau.

The phone rang one day in September; to say that an old school chum of Jill's named Betty Dobbin was coming over from Canada and wished to meet up with her again after many years. So, it came to be, at Ullesthorpe Court on the golf course, with another old school friend Angela and husband Frank. This was to be the start of several annual meetings at various pubs locally, until Jill became less able after about 2012. Shame.

Cassini arrives at Saturn in October this year and goes into orbit, but it will be early 2005 before the Huygens probe is launched into Titan's atmosphere, more later. And finally, my 70th birthday arrives 18 November, where a good time was had by all, here at 47 Long Street.

2005. Success, the Huygens probe successfully landed and sent back amazing pictures of the landscape in March of this year. Just after this we heard that there would be a seminar/lecture at the Leicester National Space Centre and having initially worked on this project in its feasibility stage, we decided to attend. It was superb, as you might expect, given by some of the people who were responsible for one of the most technological achievements to be seen in our lifetime – what a privilege. I was also able to ascertain that the parachutes used were developed by Irving's Limited, now owned by Hunting's.

PS. I have the book covering all this, it is called *The Titans of Saturn* by Brian Groen and Charles Hampton- Turner. Well worth a read! Also, on the strength of all this, I wrote to the BBC suggesting that they show some of the old Will Hay films, as a tribute to him as an astronomer, specialising on Saturn; as most of the general public are not aware of the fact, but apparently, he made the films to pay for his prime interest – astronomy. They declined – too much to hope for, I suppose.

We continued with constructing the E3 for what was to prove its final demise after disappointing results, as follows: The final timings selected – ref book, gave an exhaust closure of seven degree's ABDC which was satisfactory and assisted by Brian Herbert we installed it in the frame. The first test run was 14 miles around my test route with Herbert as a backup and it was immediately obvious that it completely lacked power, in fact into wind or up a slight gradient, it would not pull in top gear! It also ran very hot, such that the thermal switch controlling radiator temperature never went off.

We fiddled with it to make sure that nothing else was causing the problem i.e. ignition timing slipped or carburation etc but after a total of 50 miles including a trip to the BSA owners at Walton, we called it a day! However, on the bright side, the fuel consumption was brilliant at 84 miles per gallon at an average speed of 40 mph – a record I should think for a 500cc two stroke engine. Our computer program had predicted this as 4.5 brake horsepower which caused much amusement to Pepper as you would expect. Never mind – they say you learn more from failure than success! Quote by H. Ricardo.

PS. The idea could be resurrected using phased transfer system to advantage – later notes will comment on this.

So, it will be interesting to see what Roger Cramps desmo engine with similar timings achieves! After this the E4 was refitted and continued to be ridden throughout the first half of 2005 with visits to Fleckney, Foxton Locks, Market Bosworth and Roger Cramps, etc. When it was stripped down to have variable timing fitted which was first used on this engine in early 2006.The E5 meanwhile continued with its manufacture until the second half of the year – July. Whence it was fitted into the frame for test runs. this was relatively easy as the bottom half of the engine was common to both E4 and E3, so we only had to pop on the new top half with its model aircraft type, radial porting – suggested by Gordon Cornell – 'Pepper' to you. This, of course, like the E4 had the rapid port openings and counterbalanced arrangement, of the reverse sleeve. The first test run was in July, over my usual test route, followed by more runs; another visit to VMCC meeting at Stanford Hall with Jack, J. and M. Lane and Derek; as before much interest was shown and the bike behaved itself. It went on until October, to record a total of 252 miles, before strip down for variable timing.

This requires some explanation, as it is not the same as the Epicyclic Gears of the E3, but involved the removal of the crankshaft timing gear and an external lever fitted onto the sleeve shaft, which was linked via rods to a hand lever mounted on the frame, operated by the rider, this assembly was supplied by Brian Herbert, left over from a pre-war Triumph he once owned! So, the sleeve in effect becomes a movable liner and by raising or lowering it, the ports move up or down, changing the symmetrical timing. It first ran in October, over the usual test route to Broughton Astley etc and subsequent runs until the end of the year.

Very little difference in performance was observed with this type of porting, so the project was brought to an end in December, with the cylinder barrel being modified to fit the new two stroke diesel project, E7, scheduled for later. Our hopes now rested on the same trials on E4 in January 2006. We also now started in July on the E6, which was a return to uniflow sleeve valve, only this time driven by a variable timing, eccentric con-rod; this at a stroke removing the sleeve shaft, it's driving gears and oil filled timing cover, so that everything could now run on petrol only! Great. It meant that the sleeve would have no rotative movement that Ricardo seemed to think essential for lubrication, but we figured that in a crank case full of petrol this would not matter significantly. Time will tell! Read on to 2006.

The usual three arrived at Lochboisdale (scene of some of the worst clearances) In April this year, to the same bungalow we rented with Charlotte and the Tebbuts in 2002.We occupied ourselves with local walks along Milton beach and made an abortive sortie to attempt the climb of Ben More at 2,015 feet but were defeated by boggy terrain, a series of small but awkward waterfalls due to excessive previous rainfall and snow on the upper reaches. But we did climb Ruabhal on Benbecula again and revisited Loch Druidbeg, Grimsay near Loch Barg Mor and Buaide Rairnis on Benbecular where a kind lady farmer took pity on us and gave us some free-range eggs! Now we come to the exciting bit! The arrival one extremely windy day of Miles and John Irwin and on bicycles; they had to pedal downhill! We knew they were coming of course, as they were over here to do some rocket firings into the Atlantic, from the Geirinis firing range on South Uist. As a result of all this, they gave us the dates and times of the proposed firings and we took ourselves off to a convenient high point near 'Our Lady of the Isles' statue where we were able to watch and photograph the firing; whilst Jack timed the flight on his wristwatch, from launch to splash down in the sea, I viewed through binoculars.

Armed as we now were with the time of flight, the angle of launch from the photographs, I was able to calculate the altitude, range of the projectile plus initial launch velocity when I got back home, using my applied mechanics book – it's all in there on pages 34 to 35 under Ballistics! (And ref our photos.) The best firing, we saw from the top of Hecla, was an altitude of 6.16 miles and a range of 17.44 miles, very impressive, lovely smoke trail for starters. These details – Miles tells me, are secret but as you can see it doesn't take much to work out – so the Russians won't find it too difficult either!

In order to get to Hecla at 1,969 feet, we started from Loch Sgioport on the eastern coast which offered the least climbing distance but for some reason Jack did not come with us. It took us a long time over rough terrain, where we saw a herd of deer and much wildlife, before we reached the summit where I am afraid, I was taken short! Most embarrassing, I was on view to the whole world also it seemed – don't ask what Jill did, there was no cover (ref photos)!! Our descent back again was very slow, it was tough going and about two miles of it; such that it was early evening by the time we had motored back to the digs to find Jack in an agitated state, about to phone for help. We shouldn't have left him so long, no mobile phones! After all this excitement, life returned to 'normal' with the departures of Miles and John Irwin who apparently was not impressed with this 'wilderness' as he called it.

We continued our holiday into the third week by going further north into North Uist past Eaval to look at Crogearraidh at 585 feet, a little 'pap' of a hill, that Jack sketched; lovely old cottages at Malacleit and as far North as Berneray to view the mountains of Harris, before returning home as usual via Lochboisdale and Oban. Soon after all this we were invited to visit the Tebbuts in their new home in Ivegill in the north of the Lake District, where he showed me his proposed new workshop in one of the old barns that went with the property: While Sandra gave Jill and Jack a conducted tour of the extensive garden.

They lived at the beginning of the private road, next the lodge house, leading up to High Head Castle, which was now a burnt-out ruin, set on fire some 150 years ago by a careless servant with a candle! We toured the usual local area visiting Cockermouth for antiques and books and as far north as Skiddaw for some reason? The pub at Ireby proved interesting, where we went for lunch, they brewed their own beer and it was run by a syndicate of local people who had bought shares in it. Good on them! In June, Jill, John, Lyn and myself were invited to Barry Roaches 70th birthday party, to be held in fancy dress at the Coulston Court Hotel – Golf Club where my father used to play with the Rushes every week, during his retirement. (Ref photos) it was a good do, involving a traditional jazz band and we all had a stomping good time.

We were just about recovering from all this when a phone call from the Lanes asking if we would like to join them in house sitting for Ron and Monty in Kinnerton. Apparently, they were off on holiday and for safety sake, asked John and Lyn if they wanted a cheap holiday. So off we went again, our first visit had been in 2002, if you remember.

This time we went further afield, visiting Madogs Wells to see Michael and Anne plus the countryside of course and an extended visit to the Tregaron area of outstanding beauty in the most glorious weather (ref photos), where we took our own food.

We followed this up with a visit to the National Trust gardens on the River Wye, South of Hay, it helped being members at that time and a second climb in this area of Radnor forest, of Great Rhos at 2,145 feet, which is the highest hereabouts and about a five-mile walk; we were fit then, all in our 60s and 70s, just! Perhaps our most interesting visit was to the Hay Bluffs where we had already done quite a bit of walking and on to Clyro on the other side of the Wye Valley, where Rod Tindell's brother had a farm with a beautiful view overlooking the Bluffs. He was adapting the farm to attract tourists to self-cater in chalet bungalows, during the summer season. The Tindell's were also there, so a good time was had by all. The holiday ended on a bit of a sour note when Aunti Lyn got all bossy and told us we should be doing the hoovering whilst the Formula One Grand Prix race was on – we left a day early, so missed seeing Ron and Monty on their return.

P.S. We used to have these periodical upsets, Lyn and I but have now mellowed with age!

27 August. Visit to Swingate Farm for John and Lyns' 40th wedding anniversary, attended by Jacqueline and Richard who stayed locally in a converted Oast House; Mick and Dot, the Heasman's, Tom and Sylvia, Miles and Pat, Nick and Heather, the Durrant's, Uncle Tom Cobley and all! Followed by a visit the next day to the Durrant's where we were treated to the sight of his 'Highness' almost in the nude! Horrible! Not the sort of thing you would expect from a Whitgiftian – or is it? I told him it wouldn't do him any good! He wouldn't listen!

I had a letter and photograph sent to me by John Martin, the man who had brought my 1957 model 19 Norton that I used for so many years to go to work etc. He had sold it on, after many years of restoration and a Colin Blundell has now finished it and made a superb job of it, almost like new (Ref photos). This was followed by a trip up to Buxton to meet Paul's future in-laws, Margaret and Frank prior to his wedding in December.

Our final holiday this year was in September, when we visited Mull again with Jack and Jean, her second holiday with us. We did the usual crossing from Oban to Craigmore, having a ride on the miniature railway in the process and went back to Franchadil

336

again. It was part of a working farm, so there was plenty to keep us interested, especially Jack with his sketching. We did some local walking with Jean but failed to climb any significant hills, partly because the weather was a trifle inclement and several days were wasted indoors.

However, we did manage to get out and visit Loch Scridain, Loch Buie and Carsaig in the south of the island where the weather was a bit more cheerful. Our visit to Tobermory provided some light relief when, as we were walking along the sea front around midday, I saw a notice saying 'Reduced cost lunches available for OAP's upstairs', so taking the initiative and also feeling peckish, I nipped upstairs to find that it was for some club or other but managed to talk them into it; as we were starving OAP's – well all except Jean, who was a trifle embarrassed – mind you it was a tasty lunch! On another trip to Salen on the east coast we were enthralled to watch a group of otters fighting for whatever, which went on for about half an hour. On our way back we called in on the Tebbuts at their home in Ivegill, overnight, only to have to return after a mile or so to collect Jean's glasses which she had left behind!

In November, Ian Vernon visited Headley Cox in San Antonio when on holiday out there and brought back a selection of photographs of his workshop etc, and the progress on his latest sleeve valve engine. This was followed by Jills 70th birthday party here at 47 Long Street, where all had a good time. And in December, to round off the year, Paul married Helen at the registry office in Manchester with the reception being held at Manchester Metropolitan University. We travelled up to Hadfield and stayed with Jacqueline and Richard overnight before the do. The reception proved noisy for me and I left on my own, early and walked back to the hotel, where Jill joined me later for sleep, or so we thought, only to be woken up later by the fire alarm which resulted in all of us assembling in the foyer outside. it turned out to be a false alarm, triggered by some silly drunken bugger from another do, Fortunately not one of ours!

2006. The E4 variable timing trials commenced in January and worked without any real problems for approximately 120 miles in total, making a total of 723 miles with this engine. The timing variations were more noticeable than on the E5 and would respond to an automatic system, with servo motor that was interconnected to throttle, rpm, and fuel metering etc for better results. (Ref book E4) as I have mentioned before. Finally, cylinder, sleeve, timing cover and sleeve drive gear assembly, etc, were assembled and dispatched to Cranfield University in November for frictional tests and has never been seen again! Not that this mattered much as we were now back to uniflow Scavenge with the E6 and the two-stroke diesel E7. Meanwhile, work continued on the E6, which uses Fred's variable Vernier timing arrangements, which I have never really understood! Too clever for me! Interrupted occasionally by work on the E2 in preparation for further testing at Cranfield University in 2007.

Jill's old nursing colleague, Elizabeth Grey (Pat) died early this year, we used to visit her and her husband at Farmborough near Bath, occasionally. She was very religious, and I liked her very much, nice lady. Just before she died, she had a book called *Inspirational Nuggets* published, which I have read several times but it's extremely biased in favour of the Christian religion and typically intolerant of any other! Nevertheless, she took it all very seriously and was devoted to it; good for her, if you are going to do something, then do it wholeheartedly!

Our first holiday this year was back to Michaels at Madogs Wells with Jack and Charlotte in February and lots of snow! In fact, on one occasion when we wanted to go out, the lane was blocked by a snow plough and we had to reverse back quite a way. We did the usual visits too Pistyll Rhaeadr waterfall and climbed Rodney's Pillar, near Welshpool for the second time; we had already done the volcano in the gale, a few years before with John and Annie Uwin. On a walk down the Montgomery canal, I came across a chap who had a comprehensive workshop, so we exchanged a few ideas. On our way home, after a week we visited Ironbridge, which is to be recommended – extremely interesting place – start of the industrial revolution! (Ref photos).

At this time my daughter, Sarah had twins – one of each, Alex and Millie, so this was a busy time. Also Derek Keep of Vauxhall's, wrapped up the UFO society after 35 years; amazingly, it had lasted for so long as everybody had now left Vauxhalls and were either working elsewhere or retired. Next up, the Lake District; we had a call from the Tebbuts who wanted some house sitters, as they wanted to go to France to visit his ex-wife.

So, the faithful three, loaded up my newly acquired trailer with the Rudge and Jacks scooter and set sail via the Greta Bridge Hotel for Ivegill. Upon arrival, we procured the keys from an American lady next door and set up home. During our stay here we were visited by Jacqueline and Richard who had rented a self-catering barn conversion at Ireby, which we also visited on one or two occasions, also climbing Binsey Hill at 1,452 feet nearby. We were here for two weeks and spent a lot of the time riding around on the Rudge, with Jack on his scooter, visiting the beauty spots etc. We had another look at Blencathra where we had stayed with the Lanes, in 1997. Silloth, on the northwest coast and Coldbeck in the north lakes. We had a visit from Norman and Edna whom we met at Buttermere, and another get together, when the Tebbuts returned from France.

There was a party at Jack's house in June when John Irwin turned up with Annie and his newly acquired Morgan three-wheeler with an Anzani engine and Bob and Anna Ward turned up in his recently restored, partly by John Irwin, Rolls Royce! (Ref photos).

I also picked up Tom's 1949 ,600cc Scott flying squirrel on my new trailer from Ian Pearce in Bridgnorth, who had been 'doing it up' for about five or six years! And now claimed to have finished it; returning it to him in stages via Stoney Stanton where we added Nancy's front door for the Lane's outhouse; and John's 70th birthday party, plus the Aylesford VMCC rally, which included a ride to Tenterden railway station, etc. Tom finally saw it when John rode it down his front drive, billowing the usual start-up smoke!

Another of Jill's old nursing colleagues, Margaret Marriott died of cancer in July; we used to visit them often with Jack as I have recounted before, for painting together. I found her difficult to get on with, she was very argumentative, like me, I suppose!

Now, we are going abroad again, to the Isle of Man, for the Manx Grand Prix where Steve Linsdell is riding a Paton 500cc four stroke twin of Italian design and manufacture. To explain, Steve had for some time now, been developing this machine for an Italian millionaire, who, as I understood it, wanted to manufacture and sell these machines but needed to show their potential; what better than racing success! We arrived via Liverpool of course and went south to Derbyhaven where Steve had rented King Williams College for two weeks, for all of us. It was a public school, closed for the summer holidays but available to rent the facilities, like kitchen, reading rooms and of course dormitories – choose your own bed – plenty of room. All in all, including the Italian team, there must have been about twenty of us!

We split into groups for cooking – Jill looking after Jack and me but all eating around about the same time. We watched most of the practise from various vantage points and we're thrilled with his second place in the senior race and breaking the lap record in the progress! For which the Italian millionaire presented him with a gorgeous 500cc Motor Guzzi, (Ref photos). Well deserved! Whilst all this was going on, we three toured around the island; it was Jill's first visit, climbing the South Barrule 1,570 feet in the process and visiting the Calf of Man, Fleshwick Bay – good pub food, and revisiting our old digs in Peel, that we stayed at in 1968 with baby Miles, Lyn, John and the Symes's. We also had a trip up the electric railway from Laxey to Snaefell at 2,015 feet; we cheated this time and as a reward there was 'Mist on the Mountain', hence no view. Justice was seen to be done!

Finally, Jack wanted to visit Jeffrey Quill's grave; he was the man responsible for the development of the Spitfire at Castle Bromwich during the war. We found it at Andreas in the north of the island with beautiful views of Snaefell in the background. A great holiday, thanks Steve! Our last holiday this year took us once again to the Preseli mountains in South Wales at Fiona's self-catering establishment. We went this time with Ted and Myra Snook for a change; our first and last holiday with them, no problems, it just happened that way. The weather was good, so we took a coastal walk along the Pembrokeshire coastal path from Newport to Dinas Head and Fishguard Old Harbour, with scenic views all the way. We also went to see a mini-Stonehenge called Pentre Ifan, very interesting, I forgot what the information plaque said about it.

It goes without saying of course that we had to climb Carningli at 1,010 feet, just outside of Newport; we got up with a struggle, it's extremely rocky and both Ted and Myra were not used to this sort of thing – however we all managed it, (Ref photos).

Inspired by success we decided to climb the highest point, which is Foel Cwmcerwyn at 1,742 feet, that is three of us decided, Ted had had enough and thought he would get a bus to Newport and have a look around. So, we three went, did the climb, not very difficult, a lot easier than Carningli, wandered about ending up at Pontfaen tea rooms in the afternoon, then home for the evening meal – or so we thought – no sign of Ted? We waited until the last bus had passed, then decided to go and find him! Jill and I set out in our car, the only one we had brought, WEAK now coming to the end of its time with us, over 200,000 miles on the clock! Best car we ever had in my opinion! Diesel. As we approached Newport, we saw a staggering figure, lurching along the side of the road – drunk. The story he told was of meeting a man in a pub, engaging in conversation – and boy can Ted talk; and finally, missing the last bus and the rest is history! I don't think he and Myra spoke to one another for a day or two, still that was not uncommon, so I understand. Shame because they are both nice people.

Final trip was to Druidston, on the west coast, this is apparently the place where the blue stones were mined for the building of Stonehenge. These Pembrokeshire stones match those found on Salisbury Plain and the theory is that they were mined here and floated by rafts, round the coast up the Seven Estuary, then dragged over land to where they are now! Incredible, but who am I to argue!

To end the year, I was asked to give a lecture to the Taverners section of the vintage motorcycle club, VMCC by a Mr Ward who lived close by and was a keen member. The subject, surprisingly, was to be on sleeve valves! The Lanes were visiting at this time, so John came with me, we got there eventually after crossing a main road or two without stopping; ward didn't seem to notice these things! The lecture went off quite well and, in the questions, afterwards I was introduced to Roger Moss, who could be called Mr Scott! During the ensuing conversation about sleeve valve two strokes, naturally, I introduced the subject of 'phased transfer' Using Scott barrels and heads etc, and he seemed interested but knew nothing about it. This was not surprising as I had not told anybody, except close friends, for fear of possible patent right problems. So, the following day, we went in that direction and I dropped off my paper on the subject; that I had written, if you remember in 1978!

2007. We had a visit to Monks Kirby, the local village, to meet my cousin Gill, with a G and her husband Peter who had lived in Leamington Spa for many years, she was the oldest of all us cousins, from the Wallington area. We also took possession of our new car, another diesel which was to go onto record about 150,000 miles, not quite as good as the last one, X463 LUT. The first Test run of the new E6 took place in March in the presence of Brian Herbert but ended prematurely after about 1½ miles with a seizure.

Investigation revealed that a lacquer coating used in the casting process had come off and affectively sandblasted the whole of the inside of the engine. Mr Pearson, the man who supplied the piston apologised and said he was not using it anymore! However, a second test run in July went well, as did several others and the BSA owners rally etc, the machine running well for the rest of that year. This engine was the best of all these uniflow 500cc two strokes, more power, less fuel and greater reliability than the previous marks! And also did a greater number of miles, 2167 in total.

Tragically, John Syme's died about this time, as I have mentioned, he was 75 and I found his loss disheartening as I really enjoyed his regular visits; he left a wife and two daughters, one Belinda who was suffering from multiple sclerosis and being nursed at home by John and Anne; I keep in contact with her, usually at Christmas.

Our first real holiday this year was back to Dalton Hall with Jack, whilst our two friends Charlotte and Jean stayed in the village of Dalton in a bed and breakfast. Local trips to Richmond's Millgate house gardens and nearby waterfalls were much appreciated by the gardeners! As were Wholeton Bridge and Barnard Castle. Unfortunately, during a visit to Barningham and Egglestone Abbey, Jack fell over and damaged his camera, Fortunately not himself this time!

This seems to be a holiday of waterfalls, we visited at least four. Next up was the Asygarth Falls, followed by the Hardraw Falls near Hawes, where you have to go through the Green Dragon pub and pay! We also revisited Tan Hill, it was Charlotte and Jean's first visit, and later a trip North to Highforce waterfall near Middleton-in-Teesdale also on the Pennine way, where we lost them both and had to return alone in our car, via Brough on the A66.

Work continued on the E2 preparing it for research by Cranfield University. This was more applicable for them as they were primarily concerned with four stroke development and had only taken the E1, two stroke because it was available at that time. As we had the crankshaft tied up in the E6, a new one had to be made and this was done at Mahle's expense by their subcontractors to my drawings and the conrod was made by Arrow Precision on Hinckley Industrial Estate, locally. The entire assembly was 'beefed up' so that thinking ahead, it could be used in the E7 diesel when it became available.

The E2 engine was finally completed and collected by Fred on Monday 16 April for delivery to Cranfield but he stopped here long enough for a midday meal at The Bluebell where the clientele were amazed at his immaculate appearance for a man who was at work, i.e. matching suit, polished shoes, gold cufflinks and fob watch and chain! But that was Fred, always beautifully turned out for work.

We took another brief holiday with Phyllis to Baildon; land of her birth, hoping to repeat the experience I had with my father in 1988, we also stayed at a bed and breakfast in Ilkley. We toured the area, visiting Sandals Road School, our home at 30 Ferncliffe Drive, Baildon centre and Shipley Glen etc, but she showed little interest and seemed not to remember much. All became evident when we returned home after a few days, when we received a letter from the bed and breakfast hotel asking for their bed linen to be returned or recompensed! We were astonished, as you might well understand but investigation revealed that Phyllis had packed it up with her luggage and taken it home, the poor girl was suffering from the onset of Alzheimer's disease and sure enough the following year she went into care and we sent a cheque to cover the damages, with explanation.

After this, the E1 reports arrived from Cranfield and as I have mentioned before, gave a very good account of all parameters. (Ref reports for full details.)

Walton-on-the-Naze features next, where we went on a visit to see another of Jill's friends who had worked with her first husband Ray during his time with maladjusted boys. She lived on her own, a divorcee, anti-men I felt; Was in some kind of cult – whether religious or not, I was unable to find out and her name at this time was Gloria, later to become Ria! Very odd. She hosted us well and we went with her to a local sea shanty show where she joined in, singing old sea shanty songs and playing the banjo, very good they were too! We also went to Ipswich, Aldborough and Pin Mill.

Back to motorcycling, not the Three Tors this time, but the Banbury run, a VMCC event done by John in 1957 which he now wanted to repeat as a celebration, 50 years on, on the same bike (ref photos). So, he and Derek arrived on their bikes for a long weekend at Stanton, calling in on Jack at Flitton on the way. On the Sunday morning, of the day Derek's Triumph refused to run on both cylinders; it had, it transpired, been misfiring on the way up from Kent. So, John not wanting to miss the event, set off with me on the Rudge, leaving Derek in Jill's fair hands to come on later when the trouble was fixed! We also left instructions of the route to follow.

However, the best laid plans etc, we arrived and John was admitted on the strength of his pass etc, which of course I did not have, not being a competitor as such, but all was not lost, the man on the gate took one look at the Rudge and waved me on – it transpired later that anyone with girder forks was let in free! A good day was had by nearly all, ref; John completed the run, I met lots of people from yesteryear, including Simon Grigson, John Hurlstone Archie Beggs – all ex-vintage races and Titch Allen who sold me a signed copy of his memoirs book, called 'Titch'.

We returned home to find a disgruntled Derek, who it appears, successfully got the bike running OK, set off and ended up in Coventry he said. He actually ended up in Rugby, although he wouldn't have it; when he failed to stay on the Fosse Way, as he had been instructed!

Work now started on the E7 in July; as you remember this was to be a diesel 500cc, two stroke based on the E5 radial ported sleeve and barrel assembly grafted onto new E3 type crank cases with the, also new, strengthened crankshaft from Mahle, and a piston supplied once again by Mr Pearson from Solihull; it continued for the rest of the year. Talk about a Meccano set, we were swapping bits from one to the other and somehow it all seemed to work, well most of it anyhow, crazy!

We had a visit from Jill's old friend Freda and her husband from Australia and also another visit from Betty Dobin her old school chum from Canada. We also went up to Manchester for a baby naming ceremony, instead of the conventional christening, for Sadie – Paul's first child born last December and stayed a couple of nights with Jacqueline and Richard in their Hadfield home.

Now it was back to the Isle of man again only this time with John Lane who came up to Leicester by coach for one pound each way – incredible. We went via Liverpool as before and had to have one night at a venture centre near Ramsey, as our digs at King William college were not available. These comprised of horse-riding stables, converted into self-catering units, rather like being in a youth hostel, very small for the three of us. In the evening for something to do, we walked down the railway line to Ramsey for fish and chips, evening meal.

The days were spent much as before watching Steve and Merv fettling the Paton and Olly's 400CC Yamaha with which he won the newcomers event on, by a considerable margin, thus denying several following riders – silver replicas, such is the exuberance of youth! Steve, on the Paton finished third – 1.8 seconds behind second man and closing fast! Our days were spent touring the island, visiting Niarbyl Bay, Port Soderick, Laxey and watching the practise and later the racing. The only incident of note was when John fell out of a pub in Castletown and broke his credit card in half!

I went to the Scott rally at Stanford Hall, met up again with Roger Moss, the result being that he liked the idea of phased transfer and would be happy to help, provided it was done in conjunction with the Scott owners and the results published in their bi-monthly magazine; *Yowl*; so it came to pass.

In September, we visited Mire Garth in the Deepdale Valley for the first time with Jack, Jacqueline and Richard. The digs were an old stone-built farmhouse called Garthmore and we met them there with their own car, a Ford Fiesta. It was a steep drop down to the bottom of the valley from the road between Dent and Ingleton but at the turning point, there was a beautiful little waterfall and pond that Jack sketched several times, over several visits – it was to become a regular holiday home!

During our time here 'Project Richard' was evolved and implemented. It involved getting Richard to the top of Pen-Y-Ghent at 2,252 feet, as this was the only one of the highest three peaks in Yorkshire that he had not climbed with his father, together they had done Ingleborough at 2,349 feet and Whernside at 2,392 feet, they are all visible, in fair weather from one another. We set off for Horton-In-Ribblesdale and the weather deteriorated, but unabashed we, that is Jill, Richard and me, set off up the climb which is also part of the Pennine way, leaving Jacqueline and Jack in the car as observers! The climb was steep in parts, but we made the summit triangulation point, just visible in the mist (ref photos). This meant that Richard finally completed the three summits in 25 years, surely a reverse sort of record, in itself! He also repeated his ascent of Whernside with us, from the Ingleton/Dent Road where we were able to obtain good views all round and were able to see the Blackpool Tower in the west!

A visit from Jill's old friend Valerie Morris from Biggin Hill was next, she stayed a few days which included a trip to Foxton Locks.

As the E6 was going so well, we thought an article in the motorcycling press might be in order, so Mick Duckworth of *Classic Bike* was contacted, he had done some articles for Steve in the past. Under the title of 'The Shed Issue' a series of 'Men in

Sheds' was written, including ours, of various individuals creating motorcycles in their shed type workshops, ours was Jill's old pigsties!

He rode the bike up and down Pingle Lane where Jill's mother used to ride her bicycle in days gone by, and was reasonably impressed; more, I think with the conditions under which Fred and I worked! He also reviewed some of the previous engines we had made and was interested in further articles on the phased transfer E9, that he had got wind of but I suppressed that one for reasons given previously but it all went pear shaped when the Scott magazine started publishing their reports later in *Yowl*!

Phyllis came for Christmas as usual with visits from Miles and Pat, Richard and Jacqueline on Boxing Day. Miles and I went on a Boxing Day run. Miles, for the first time, on the Rudge and myself on the E6. Upon our return he said how impressed he was with the Rudge, expressing his delight at the 'grunt' it had, I think it has revised his ideas of taking it on someday! Off to the Lanes for the New Year!

2008. The drawings of the E9 phase transfer Scott engine were started in January and continued for a couple of months or so before completion. Following this, copies were sent to a pattern maker in Melton Mowbray and followed up by castings being produced by a foundry in Coalville, ready for Roger Moss to start initial machining early in 2009. Whilst Fred and I started work on the other items in October.

Meanwhile we paid a visit to Tom and Sylvia who had just moved into a large, four storied house in Herne Bay. totally unsuitable for them as it had for lots of stairs from cellar to attic and the loo was on the first floor upstairs, both were in poor health approaching their 80s. It was not surprising that a few months later Tom died; he had been a Scott enthusiast for many years and left a lot of Scott information and bound magazines etc, which subsequently I passed on to the Scott club.

Jill had been feeling a trifle unwell for some time and went for a biopsy in April, but nothing untoward was found. Later she became ill again in Bowland but whether there was any connection or not I do not know!

Bad news from Cranfield was received they had managed to seize the engine E2 and tore the foot off the sleeve, broke the piston etc. I suspect inadequate cooling, particularly in the junk head region, so suggested bigger fans next time! Sleeve was blue with heat! We now had this to repair in addition to all our other projects, just what we didn't need! There was an inevitable delay as we didn't have another sleeve, piston or crankshaft and drawings were sent out again for replacements for these items. Later in the year, the engine was again returned to them, with our fingers crossed! Remember the 125cc E1.

Our first holiday this year was a trip to the forest of Bowland again, with Jack; followed by a second week with a visit from Jacqueline and Richard. We three, set off for our digs in Newton-in-Bowland in May and on arrival Jill said she felt unwell and thought the car exhaust fumes might be the cause, we had a new exhaust fitted at the local garage later. However, she felt worse the next day, Sunday, so we took her to the nearest emergency clinic in Clitheroe, where they examined her and prescribed some medicine, which seemed to work as she picked up soon after. Jacqueline and Richard arrived, a couple of days later. They stayed to the end of the first week doing local walks, exploring an old mine near Whitewell and a visit to Pendle Hill.

Then we all started north for Mire Garth, passing over the Bowland Knotts, which is the high ground leading over Great Harlow at 1,586 feet, which we climbed from the already High Road; cheats! From here, we passed an interesting 'erratic' near High Bentham, which had climbing steps cut in it, (ref photos) and had glorious views of Ingleborough in the far distance stopping at Chapel Le Dale for a roadside lunch, before arriving at Mire Garth in the late afternoon. An 'erratic' is a boulder left over from the glaciers of the last Ice Age, 10,000 years ago. We explored the local and not so local area extensively over the next week, discovering a lovely picnic site, by a stream in what we called 'Lows Gully', which was in the news at this time. Somewhere in Borneo at this time, a group of British soldiers on a training exercise became lost and trapped in a narrow defile known as Lows Gully. They were later rescued, after much effort.

One fine day, Richard and I decided to climb Whernside directly from the back garden and after what seemed a near vertical ascent, we reached the upper ridge and walked south until we arrived at the triangulation point. What a view in all directions. Further excursions included trips to Sedbergh via Old Hutton's scenic bridge, for books. It is a great book centre, well worth a visit if you're that way inclined. Later trips to the Howgills and Courtly Spout, plus Dent Station and beyond to Garsdale Head.

In July, John and Derek visited for the annual trip to Stanford Hall VMCC meeting where Derek continued his search for AJS bits and we returned the compliment in August by our annual trip to VMCC Aylesford rally, where Derek 'continued his search for AJS bits'. After some 987 miles the E6 decided to rattle and lose power and subsequent strip down revealed that water had got into the combustion chamber via the junk head 'O' ring breaking down. We replaced this with a Viton one and had no further trouble from this source. At the same time, adding further counter weighting to the sleeve eccentric and reducing the water volume in the barrel, thus speeding up the flow rate around the hot exhaust region and other minor jobs. The machine was up and running again by October and continued to perform satisfactorily until almost the end of 2009.

At this time, I was invited to give a lecture to the Scott owners at their annual rally at Abbotsholme School, where Alfred Scott was educated, to present my concept of a 'different variant of two stroke engine', Roger Moss's words in the *Yowl* magazine! i.e. 'Phased Transfer'.

Armed with the GA – General Arrangement drawing and a cardboard model of the transfer port system, I endeavoured to explain how it all worked, I am not sure that a large section of the audience quite grasped it but later articles in *Yowl* – the Scott bi-monthly magazine, I hope made it all clear! The lecture was recorded on video camera by Richard Moss and various offers of help were obtained.

Another engine came into being this year, the E8, which was a derivative of the most successful E6, to date. Basically, it used the eccentric drive, reciprocating sleeve as before, but due to its over square dimensions, the entire drive mechanism was located within the bore diameter (ref drawings). With a bore of 100 millimetre and stroke of 64 millimetres, it was designed for high revs! Over squareness does not favour the barrel ported two stroke, as the port to wall area decreases as the stroke/bore ratio drops below unity, however this is restored by using the sleeve valve with its fully circumferential porting. The four stroke of course benefits because more or bigger valves can be accommodated within the larger bore.

Two types of timing will be available, symmetrical for use with expansion chamber and high power or asymmetrical, for economy with early closing exhaust ref E3 and/or phased transfer!

Unfortunately, it remains a dream (ref drawing E8). The E7 diesel which we had also been working on all year was now finished and ready to go in the frame, at the moment occupied by our E6. So, it was put on the bench and has remained there ever since due to the pressure of building a complete Norton motorcycle for the E9; at 75 years old, creeping old age, looking after Jill, and eventually the death of Fred, have slowed output, however, I have hopes that somebody will perhaps take it on!

Our second holiday was to the deep south for a change, Dartmoor with Jack, Charlotte and Jean. We stayed at a farm self-catering barn conversion, except Jean, who for some reason or another booked a hotel in Chagford. One of our reasons for choosing this place was because Jack, at some time, had a friend living in this region and wished to find the grave in Chagford churchyard. Unfortunately – wait for it – he couldn't remember his name! Typical. Never mind it was a nice setting and there was a pub nearby. While we were here, Jill and I climbed Buttern Hill at 1,345 feet, which had a triangulation point on it, so all was not lost.

Fanworthy reservoir, close by was visited as was Meldon reservoir, which we walked around whilst Jack sketched and a trip to Widecombe in the Moor for afternoon tea. We also went to see Paul, Helen and Sadie in Dawlish, of course, and on the way back spoke to a couple of cyclists who were on their way to John O'Groats from Land's End! When we returned, Jean informed us that the pub she was staying at was offering a selection of curry's; all you can eat for £5, say no more, Charlotte and I eat the most, Jack a good third but Jill and Jean put on a poor show!

Dave Tebutt appeared in October for his annual visit to the Coventry model engineering exhibition, where we met John Urwin, Miles and some of the Hunting's lads. Our second visit to Mire Garth this year, required a bit of pre planning – it went thus: Jill and I on a previous visit to Jack's house, discovered that Tommy – his nephew and his wife Pat, plus Trish – his niece and her husband Gav, were also on holiday in this area at the same time, so it was arranged that, without telling Jack, we would all meet up at Tan Hill. Meanwhile we drove north to arrive and meet Jacqueline and Richard at the farmhouse and spent the intervening time exploring Coverdale, Aysgarth and visits to Sedborough for yet more books.

Upon arriving at Tan Hill, we discovered that the others had not arrived, so we ordered drinks and seated ourselves. When they did arrive, Jack, with his back towards the door, failed to notice and Trish stepped forward, tapped him on the shoulder and said, 'Would it be alright if we took this chair, Sir?' Jack started to reply before releasing the situation, it was all very entertaining, and Jack really enjoyed seeing them! (ref photos) A nice surprise. From here we visited Kirkby Stephen which is high up in the middle of nowhere; I bet it's bleak in winter and on a return visit from here, near Nately, we came across Pendragon Castle, a very picturesque ruin perched on a small hill, with some curious little stone tunnels etc, all very interesting.

We went down to see Sarah and the children later in the year, only to find that the rear garden fence that Gareth was building to stop the children from falling over onto the river towpath, was unfinished. So, I set to and spent a few days rebuilding it, all good fun! We attended Iris's 70th birthday party at her house with the rest of the family. Aunt Bessie, who was Cousin Bernard's wife died aged 90+ in November, the service taking place at Darlington church, Northamptonshire. Unfortunately, John and Rosemary missed it, so Jill and I were the only Collett's represented! When they finally did arrive, we retired to the local pub to discuss whether or not we should look into the inheritance question, as everything, more or less, including the property was apparently left to the executors, i.e. the lady who had been caring for Bessie, her next-door neighbour! John and I had not seen a copy of the will either, I don't know what uncle Tom would have thought about it all but John was of the opinion that the lady in question had probably earned it with all the care and organising problems, so nothing further was done! I occasionally wonder what would have happened had we have challenged it.

This was also the year that Miles and Pat divorced; it took him by surprise I believe, he wasn't keeping his eye on the ball! It was also the year of the financial crash, boom and bust, that Gordon Brown seemed to have forgotten, although I had read an article in a *Telegraph* some months earlier, which I still have, predicting the very same! What hope is there? Anyway, it's all contrived by the rich, to get even richer! (Ref 2019), where we are now; we seem to have all the money in the world to squander on everything; what price 10 years of austerity! Finally, on a more cheery note, we attended a fancy-dress party for the BSA owners end of year do, only to find upon leaving late at night – snow, the start of a tougher than average winter! (See 2009.)

2009. Roger Moss starts machining crank case castings for the E9 in January at weekends only, as his 'life supporting' work has to take precedence, obviously. A second article is published in *Yowl*, the concluding part of phased transfer operation with belated diagrams vital for an understanding. An editorial 'slip' meant that several people found difficulty in following the previous text!

Another holiday to Mire Garth comes next in March, which we were told whilst there, that it was the coldest winter for 30 years and looking at the photographs you can see it! We went this time with the Lanes; their first visit, Richard and Jacqueline but no Jack, it was to be the last visit here that we were to make. Most early mornings, three of us, Jill, Lyn and me went on a walk, usually up to the waterfall and along the Ingleton/Dent Road, one way or another; the idea was to get Lyn into shape for her forthcoming West Highland Way, Loch Ness and coast to coast walks she was planning to do with Ann and Lyn Tindall in the forthcoming years! Good for her.

We did most of the usual places we had visited in previous years, like Sedbergh, the Howgills, Ribblesdale Viaduct and Dent Station with a new trip to Barbondale just over the other side of Crag Hill, at 2,230 feet which at this time was covered in snow! We also celebrated our wedding anniversary 6 March by having an evening meal at the Sun pub in Dent alongside a roaring fire; which reminds me that the Lanes burnt all of our log allocation at Mire Garth to which we had to pay extra! Mind you it was worth it!

The Easter meeting at Brands Hatch featured John Lane riding a vintage CSI Norton, as he had done in 1969 some 40 years earlier! He decided to do it following an announcement by the British Motorcycle Racing Club's – BMRC centenary meeting, asking for vintage races who had previously raced at their meetings.

We went with Jack, Miles, Derek and bike of course, no problem except leather's. He had previously tried on his own, no hope at all, just like me, I can't get into mine now, so Steve sent a set of his down, no hope either. (Ref photos.) this photograph is now on display in his shop at Flitwick for the amusement of staff and customers, how cruel! An appeal over the Tannoy produced a suitable result and John was able to ride round several laps, which he really enjoyed with everyone else.

The E6 meanwhile, was going well and I was invited along with Roger Moss to the British two stroke clubs annual meeting at Market Harborough, where we demonstrated his Scott racer and my machine; we were further entertained by a talk on Rogers life, all very amusing, he is a great recontour. I also went to the festival of 1,000 bikes at Mallory Park and was invited to display the bike on the BSA owners stand, the result being, later in the year, the machine appeared in the VMCC magazine, with a query? What is it?

Jack's final holiday with us was to the Tebbuts in the Lake District, he was now 88 and beginning to feel his age, I think he would have liked to continue but couldn't trust his waterworks and bowel functions, so regretfully, not wanting to be a nuisance, he declined. If he had explained more fully, we could have perhaps done something but that wasn't Jack, so be it! Not much to say about this one, we visited the usual sites, including the pub that brews its own beer. Dave met an interesting chap who was repairing a water pump at this pub and discovered he also had a workshop and was of like mind, I wondered since whether anything ever came of it.

Sarah's son, Leo, was born in August. A bonny baby!

Jazz came next, an invite from John Eastern to celebrate his move from Olney to Old Western in Cambridgeshire with a party that included the Tebbuts, some Hunting's workmates, including Mike

Jones who had run the Bovril test facility and played bass guitar. Very good too they were – although Dave is not a jazz fan, he prefers Sibelius.

We returned to Abbotsholme, taking with us the machined E9 crankcase castings from R. Moss and various other components that Fred and I had completed to date, plus photographs of the workshop and various machining operations, these were very well received, and many questions were asked. Fred as ever, was sceptical! Cranfield were on the phone complaining of some knocking in the engine E2, which we had repaired previously, if you remember, and thought it might be the big end! So, we said, return it, which they did. They had partially dismantled it by removing the barrel and sleeve assembly etc in an attempt to check the big end – but found nothing wrong, so they remained puzzled!

However, further investigation by us revealed damaged splines on the crankshaft and its associated drive pulley giving a very loose fit and showing plenty of fretting. This was the source of the trouble and it was due to them, trying to start the engine without a clutch, we had warned them before about this and I thought they had sorted it! However, they said they did not want it back as they had completed their test work during the year and sent the relevant papers to us, which we still have. Fred has now semi-retired, that means that he can come and go to Marhle as he pleases i.e. flexitime; all work on the E2 is now finished and it lies in a box in pieces! Awaiting whatever. An inglorious end.

Our final canal holiday comes next, with Charlotte and Jean, starting like the first, from Stoke Golding on the Ashby canal and with the same boat, the Bosworth 68 foot. There was an initial trial whilst one of the boatmen made sure that at least one of us could 'drive' the thing! Our first main problem was getting onto the Coventry canal at Marsden, it's a very tight turn with such a long boat. We stopped later near Hawksbury for a pub meal and then continued on to Brinklow for our night stop. Then disaster struck, Charlotte discovered she had left her purse behind at the pub! A mobile phone call established that it had been handed in and another call to her daughter Judith, who agreed to collect it and bring it to us, saved the day!

We left the next morning and passed through the outskirts of Rugby, heading for Braunston to get onto the Oxford canal, another tight turn, when real disaster struck. I think Jill was 'driving', Jean up front, Charlotte inside and myself at the rear with Jill. When all of a sudden, the boat was skewed across the canal and heading rapidly for the opposite bank, I leapt off and to the towpath, to try and prevent the stern colliding with the bank and in doing so slipped on the mud, bashing my ribs on the guard rail, leaving me winded. Meanwhile, Jean had picked up the pole and pushed it into the bank to try and fend off a frontal collision. the pole snapped with a sharp edge which fortunately passed to one side of Jean and went through one of the front windows.

It was a very close call, she could have been impaled, with the weight of a 10-tonne boat behind it, I have a cold sweat every time I think about it! As far as myself, I was in pain for the rest of the holiday.

We pressed on through the outskirts of Braunston, canal centre of England's waterways, and eventually arrived at Shuckburgh and Napton, where we stayed for a couple of days also getting the broken window repaired by another boat hire company, who said this was normal practice to help one another and we would be charged damages when we returned. I think it was about £50 for the pole and window! All's well that ends well. The return trip was relatively uneventful, apart from another skew across the canal when Charlotte was driving and which I apparently swore at her to let go of the tiller, so I could try to regain control in the heat of the moment. If so, and it is quite likely, I do apologise! To sum up; the crew generally were pretty hopeless at 'driving' the boat and could not be left alone for more than a few moments before some disaster or other became imminent. This meant that I was on call all of the time, hence never got a chance to stretch my legs by walking the towpath! But the cooking was good and all other duties, except cleaning out the propeller shaft box, were undertaken with diligence and care.

2010 started on a bad note, as this was the first time that I started to notice that Jill's memory was not as good as previously and she had become slower at doing things.

A further progress report of the E9 appeared in the *Yowl* in February with a graph showing the exact operating functions; thus, completely confusing everybody! Also, acknowledgements to R. Moss for his machining of the crankcases, Dave Holder from Meriden who has supplied the barrels, pistons and con-rods and Rex Caunt for the ignition system.

Our first holiday this year was to Manaton on Dartmoor with John and Lyn where a self-catering cottage had been booked and very cosy it was too! Lovely log fires (ref photos). We started a pattern of early morning walks, as we had done before at Mire Garth, to keep Lyn fit for her intended walks to come; whilst this was taking place, John was getting up and preparing the breakfast, good lad! There were some interesting rock formations at Manaton Moor and one called the Bowman's nose was a very 'life-like' face.

We also took trips out to Widecombe, Lustleigh, Fernworthy Reservoir again and climbed the summit rocks on Hey Tors at 1,475 feet. A sign of the times perhaps, is that Jill did not come with Lyn and me! Of course, whilst we were down here, we visited Paul, Helen, Sadie and Martha and took them out for lunch and a visit to the fun fair at Teignmouth, where a good time was had by all. Jill's eldest son Alan decided to get married to a Canadian girl called Sandy and live in Canada. We motored down with Jacqueline and Richard driving to Paddington basin in London, for a canal trip along the Grand Union Canal, with hog roast meal etc, staying overnight at the Premier Inn in Edgware before the return.

Earl Sterndale in the Peak District was our next venture, when Charlotte came up with the idea that she would like to do a three-hill walk, she had seen in her walking magazine. These included High Wheeldon 1,372ft, the Dragon's Back and Chrome Hill on the top of the world. We completed the first one easily, stopped for lunch at Earl Sterndale and then struggled with the Dragon's back, just Charlotte, Richard and me. It was very steep on the descent and Richard had to help Charlotte down a near vertical section, I had gone another way which wasn't much better. We ran out of time for Chrome Hill, but Charlotte vowed to return.

The E6 continued to run well but a misfire developed and was traced to worn out contact breakers so a new MZ set was fitted to cure the problem. Titch Allen, founder of the Vintage Motorcycle Club and personal friend, died in May, his memorial do was held in Kirkby Mallory Church, with the ensuing wake as a thrash around the circuit which had been hired for the occasion – brilliant. I went on the Rudge, dressed as usual! I was alongside Roger Cramp; he was riding his 1926 Cammy Velo, and on my own when the flag dropped! When all was over, I returned to the paddock to find that as he let in the clutch, his rear chain came off the rear wheel sprocket. The only other time this had happened to my memory, was on our way to the Irish rally in 1990 with Jack, when it came off the Rudge! What a shame, he missed the thrash.

All the family, or almost all came to Stanton for a weekend in August and Mick allowed us to play tennis on his top paddock. (Ref photos.) Jill and I went on our own to Madogs Wells again in June, did the Solar System walk, visited Bishops Castle and all the regular spots and went on the Llanfair to Welshpool light railway for the first time, it was great fun, well worth the effort.

We went down to the Lanes Alesford VMCC rally, which was becoming an annual event and took the E6 and the spare crank cases of the E9 with Fred's wooden cylinder barrel to form a dummy engine for fitting trials within John's Featherbed Dominator Norton frame. He had removed the engine and gearbox etc some years earlier with the idea of refurbishing it, eventually. So, it was perfect for these trials; however, he refused to sell it to me at this time, saying that he did not want it cut up as would be necessary to fit the double Scott engine by lengthening it by some six inches! Removing the E6 from the trailer, I rode it to Ightham Mote as part of the vintage run, with John on his faithful CSI Norton to meet Bob Durrant there on his 125cc Honda. The following day was the Aylesford rally, where my machine attracted the attention of the British two stroke club and it went on display on the stand where many questions were asked.

Back to Madogs Wells again in September with Charlotte, Jacqueline and Richard, where we took another ride on the steam railway to Welshpool via Castle Caereinion and later on visited Montgomery Castle, followed by a trip to Corndon Hill at 1,667 feet, where Jill and I had previously climbed some years before to see the stone circles and Mitchel's Fold monument. Richard and Jacqueline did not come on this one as they were ill! Finally, visits to Lake Verney for lunch at the hotel overlooking the lake, where I believe some of the test flying for the Dam Busters took place, returning via Pistyll Rhaeadr waterfall which was in spate. At the end of the year, I went with Brian Herbert and family to the Abbey pumping station at Leicester to help on the stall they set up every year. At this time, Brian was suffering from dementia and had given up working on or riding his motorcycles! Awful. The end of the E6 on the road now took place, after a total of 2,167 miles and due to the fact that the MOT, road tax and insurance had run out and Steve was not doing insurance from now on! It was sorned and laid up in the garage until it could be put on the road again, with the Rudge, when the new Scott E9 also became available!

Meanwhile, back at the works, the E9 engine was now finished after nearly two years work by Roger, Fred and myself. Finally, after several months of investigation we were able to purchase an ES2 Norton, 1958 vintage, with a wide line featherbed frame; the engine being useless to us, was passed onto Roger for his sidecar outfit as his own engine was 'knackered'.

2011. A further report in *Yowl* showed pictures of the completed engine with both the normal two stroke manifold's, i.e. those that send the mixture direct from crankcase to working cylinder as is done in all two strokes with crankcase compression; and the phased transfer manifold's, where the mixture is sent to the adjacent working cylinder. (Ref photos.) We had some trouble getting those sets of manifolds made; initially they were promised by a 'gentlemen' from Northern Ireland, drawings were sent etc, but nothing ever materialised, so a firm in Barwell was contacted and they sat on the project for three months, with nothing except excuses! Finally, we were put in contact with an ex-Rolls Royce sheet metal worker, a Dave Wakeling from Mountsorrel who made a superb job of them in stainless steel.

I gave a lecture to the Midland branch of the National Auto Cycle Association at the Queen Victoria pub in Syston, to an audience that included Roger Moss, Dave Wakeling, Roger Cramp and about 21 other interested parties; on a review of all our (Fred and I) projects over the last 35 years! Ending with the E9, which afterwards Roger Moss, summing up said we were witnessing an event of 'historical significance'! Well, it has yet to happen I feel; still as far as the two stroke is concerned, I suppose it is something new? Having completely stripped the Norton, we modified the frame to widen and lengthen it for the double Scott motor; jigged it all together and took it to a Mr Gary Richardson who braised it all together. By now it was June and Jacks 90th birthday, where a garden seat with his name on it was presented, we are sitting on it. (Ref photos.) John Irwin turned up in his Blackburn Engine Morgan three-Wheeler of 1926 vintage and Bob Ward appeared with his wife in his pre-war Bentley.

The Aylesford VMCC rally this year was augmented with the Foresters drive seat reunion on the Sunday afternoon of 8 August. To explain: this was an idea that Lyn, John and myself with support from Jill, had come up with a get together for the original 'mob' or as many as we could find of them after a lapse of about 60 years! We came up with 13. Prior to this, John and Lyn drove around the Wallington/Beddington area and photographed most of the relevant streets that we had all lived in at that time, and they were able to establish and sit on the original seat, now moved to the other side of the road! But still there.

Astonishing, the woodwork has obviously been restored but the concrete ends look original. (Ref photos.) Everybody had a terrific time as some of the people had not seen each other for that number of years. Lyn did a magnificent spread as usual, the weather played ball and it was so popular that another one was held in 2015.

Our first real holiday this year was delayed by a day as we attended Charlottes 70th birthday party at The Bull pub in Stanton, thus arriving at Brancaster Staithe on the North Norfolk coast, to greet the Lanes who had arrived the day before. It's a bleak sort of place with the sea always a long way out it seems, even when the tide is in, very flat. We visited Holkham Hall just down the road where Jill forgot her brolly, but it had been handed in. Hunstanton, Heacham beach and Downham Market where a book shop there supplied me with the final volume called *Wisdom's Daughter* to the *She* series by H. Rider. Haggard, that both Richard and I had been looking for, for some time! Going East was Wells-next-the-Sea, where we took a ride on the 'Wells and Walsingham light railway' which was about a three-mile ride to Walsingham across open countryside for a pint there – very nice, weather good. (Ref photos.) Walsingham is very religious and called 'The Shrine of Our Lady' but very nice and peaceful, unlike many places today. On through Blakeney Point to arrive at Sheringham to see Nick and Heather who had moved here from Cambridgeshire just recently, upon the death of Nick's father, I believe. We climbed the cliffs behind their house for great views and looking in land, were told that Beeston Bump, altitude unknown, was the highest point in Norfolk!

No sooner than we had arrived home and we were off again, this time to Cardigan Bay in Wales, a place called Cei-Bach between

New Quay and Aberaeron. We went in our car LUT and Jean and Charlotte went in Charlotte's car, I don't know why we didn't all go together! There was a problem on arrival, one of the upper bedrooms was full up with blue bottles, very nasty, so we opened the window and phoned the owner/caretaker who came and dealt with it but there were more than enough bedrooms for all so we didn't use it.

Local walks along the beach were taken frequently and on two occasions we walked first to New Quay then a longer walk Aberaeron where we had to cross a waterfall and found a field with mushrooms in, so I had mushrooms on toast cooked by Jean, nobody else would try them, although Charlotte pronounced them okay but I notice she didn't eat them!

One day, we decided to go to Aberystwyth for a ride on the steam railway to Devils Bridge, unfortunately we went for some reason in separate cars and lost Charlotte and Jean in the traffic by Aberystwyth. We waited on the platform until the train was due to depart, then gave up and went aboard. Upon our return there they were, waiting for the train; so, they went on their own – unfortunate but never mind! A trip to the Pistyll Rhaeadr waterfall was taken, followed by a paddle in the sea by myself, I like to have a paddle once a year, to wash my feet! A visit to Newcastle Emlyn to see an old woollen mill in operation, one loom being operated by an old, retired worker who cannot leave it alone!

Ted Snook came up and welded our exhaust pipe plunges on whilst in situ and took photographs etc, to send to his friendly editor of the *Classic Bike Guide* to be published in November. It gave a run down on the project E9, the Scott club involvement and rough idea of how it works, and what it hopes to achieve. All good stuff.

2012. Following the Foresters drive reunion last year, I received a phone call from Anita saying that her sister Janet was coming over from Australia in January and inviting the Lanes and us to meet her. This was accomplished and it was nice to see her for the first time in about 50 years! Since she first moved out there.

Having now got the E9 engine and frame together, I decided it was time to get it registered as I had to previously for the E2 with the Bedfordshire council in 1994.This time however things were different, the machine had a logbook and a registration number SAY 527. Norton ES2 all of which we wanted to keep – no more Q plates! so I contacted a Mr Worthington Williams c/o Old Bike Mart that I had been subscribing to for a couple of years or so and he in turn requested:

1. That we stamp the engine number shown in the logbook on the crank cases of the new engine.
2. Getting the licencing authority stamp on three logbook copies.
3. Obtain an engineers' report from an independent motorcycle specialist, in this case Supreme Motorcycles Limited of Earl Shilton which specialised in old British bikes. Armed with all this and a covering letter from Mr Williams I visited the DVLA offices in Nottingham, our nearest, to await the arrival eventually of a historic new logbook, with same registration number and a change of engine size from 498 cc to 1192 cc! Still of 1958 vintage, therefore not requiring an MOT! A great advantage, introduced in November 2011. Luckily!

Holiday wise, we went to a self-catering 'Jerusalem' cottage at Earl Sterndale in the Peak District with Lyn and John for whom it was the first time in this area; they were much impressed, always having passed it by on the way to somewhere! And the weather was brilliant all week. It was an odd sort of place with the living room upstairs and a bedroom in the attic and the kitchen in the middle, with another bedroom underneath; this was possible because it was built on the side of a hill!

Visits to the local pub were of course in order and it had a peculiar name 'Woman Without a Head' and it had just reopened. Jacqueline, Richard and Valerie, his mother visited one day and Richard, Valerie and myself climbed Chrome Hill, to record the views for Charlotte's sake; her turn will come!

Trips out to Wolfscote Dale, Wildboarclough where Jill, Ray and family used to camp and Castleton where John 'found' a vintage BSA in the car park and eventually the owner, were undertaken. We went over to the Roaches at 1,641 feet and from close by we obtained good views of the Jodrell Bank radio telescope, one of the very first big ones and used extensively in the late 50s and early 60s for keeping in touch with the Americans space programme and tracking Russian satellites!

Around this time, on a trip to the Lanes, John and I retrieved his ex-racer from the tatty outbuildings up the garden and got it into the garage where it was in the dry at last, something I had been pushing for, for years; he kept the rubbish in the dry and the valuable stuff in the wet – leaky roofs etc! Difficult to understand! Yet another visit to Abbotsholme with Miles and Ted Snook, with all the progress photographs pinned onto large boards for display in the foyer; the library was unavailable; reasonable interest was shown but time was marching on and as no running had yet taken place, peoples interest tends to lapse!

Aylesford was revisited and several carburettors and associated items from Surrey Cycles were obtained to finish the project, ready for a start-up. This was achieved for the first time on 23 August running on all four cylinders until the carburettors flooded, leading to a whole series of fuel problems, terminating in replacement 'Wassell' carbs obtained at cost by Steve's Flitwick Motorcycles, by the end of the year.

Next holiday came about from the desires of five people with four separate ideas of things they wanted to do.

First, Derek Clancy; he wanted to go back to Cornwall and visit all the old places that he used to go with Doreen and his children years ago.

Second, John; He had made contact with a chap called Robin Ord, who had agreed to buy a CSI petrol tank in kit form, i.e. solder all the separate panels back together again!

Third, Lyn and Jill, they wanted to visit the Eden garden project near Saint Austell.

And finally, Rob, who after discussion with Jack had agreed to locate and photograph the old cottage that he and the other Rob used to visit on their motorcycles just after the war. All were accomplished!

Jill and I set off in bad weather, mainly rain as usual, down the M5 motorway and onto the A30 as far as Jamaica Inn, where we had agreed to meet up with the others, between midday and 3: 00pm. We waited and waited, drank some beer, had a meal, no sign of them! So, at about 4:00 o'clock we decided to carry on alone. By now it was beginning to get dark and still raining heavily. Wonderful start we thought. Once off the A30 and the B3268, on minor roads we were lost; with no detailed map of the area, no people about to ask, visibility zero, only an address of Trebyan Forge near Prindl Pottery! Eventually, we did find someone and received directions, arriving thoroughly fed up at about 6:00pm with no sign of the Lanes. However, about half an hour later they arrived, having also had a bad journey and difficulty in finding the place. Asked about the meeting at Jamaica Inn, they said, 'we passed there about 3:00pm, late, didn't think you would still be there, so we didn't stop'. What can one say!

Our digs were converted barns belonging to Lanhydrock House estate owned by the National Trust, of which we were also members and so had free access to their house and grounds, which we availed ourselves of. We 'did' the local area including the Eden Project, which wasn't very far away and on the first reasonable day weather wise, we set off in John's car down the A30 to visit the Lizard and those places that Derek had come to see. First stop was Cadgwith Bay where the caravan site called 'Chycarne' was situated, this is where they rented a van all those years ago, not our scene at all! We actually looked for the actual caravan they stayed in – ridiculous, but that's Derek!

Then on to Lizard point, farthest point south of the British mainland where we were relieved to find and talk to a man with a Ducati – what a relief! From here we crossed the Isthmus to Mullion Cove and its hotel where they used to play on the sands. We returned via the Trelissick Ferry across the River Fal, to complete the third task and found the blacksmith's cottage on the fork of the A3078 and a minor road. Taking the necessary photographs to satisfy Jack. Excursions to Boscastle in the north and Looe in the south were carried out at Derek's request but as you can see from the photos, the weather was deteriorating. Last but not least, John's task was accomplished one evening, by a knock on the door, announcing the arrival of Robin Ord, who arrived with wife, daughter and the necessary money.

As the weather had improved for the journey home, Jill and I called in on Colin and Madge Shields at Gotherington in the Cotswolds and had a look around Prescott Hill climb and museum complex only about a mile and a half from where they lived.

2013. Off to the Isle of Wight in March with Lyn and John in their car, to sail from Southampton to Cowes and then to Braiding near Sandown for a week. There was an interesting Roman Villa nearby; the Isle of Wight airport and a visit to Sandown to see the Trouville Hotel where Iris, Miles and myself stayed in 1968 with the Hancock's, who went there every year without fail!

We went to look at Carisbrooke Castle but unfortunately it was not open this early in the year. The weather played tricks on us in the form of snow! Strangely, it struck the extreme south of England, including the Isle of Wight and even the Channel Isles, very unusual (ref photos.) It caught us out near Merstone heading for Stickworth Hall, where we got caught in a snow drift and took a downhill side road to Perreton Farm and had to reverse back for about a quarter of a mile, with me getting out frequently to bounce up and down on the bonnet to obtain some grip! Finally ending up rearwards in yet another snow drift against a hedge bank where yours truly had to get out yet again and push us out!

Foolishly, we decided the same evening to eat out at a pub on the top of Culver Cliffs on Bembridge Downs only to find that the climb up the road had not been cleared and almost all of the staff had been unable to get there, so we drank beer and ate crisps! Osborne house was the family home of Queen Victoria when on holiday in the Isle of Wight and a short distance from the chain ferry across the River Medina in Cowes; very interesting, if noisy! Along the northern coast to a wildfowl sanctuary at Newton River, past Yarmouth Castle to the Needles. Views were good as the weather began to approach normal, for this time of year.

To complete the Grand Tour of the island, we continued Eastwards along the southern coast, through Hanover Point and Niton to return via Ventnor, Shanklin and Sandown again. The trip homeward was uneventful.

A local visit each year is to the village fete at Newnham Paddox with usually Charlotte and Jean, where there is the most impressive set of wrought iron gates in the country, with the exception of those at Saint James's Palace in London, ref Lutterworth – *The Story of John Wycliffe's Town* by A.H. Dyson, of which I have a first edition, 1913!

On the home front, the garden room roof had been leaking for some time, due I think, to an insufficient angle of repose, i.e. 20 degrees or less, for slates, such that when wet, drawback occurs, water therefore getting under the slates, with subsequent disastrous Results. So, the Groococks, as usual, were consulted and suggested a new type of fibreglass construction which was duly completed.

Bad news arrived soon after, to the effect that Rosemary, John Collett's wife had died suddenly, of a massive stroke. Apparently, she had shown no symptoms of illness or else and was in good health – just goes to show! John had gone out for his morning paper with the dog and when he returned, she was lying on the kitchen floor, what a dreadful shock for the poor man. Jill and I attended the funeral, of course and for the first time in many years re-established contact with my other male cousin Martin Rush, on my father's side of the family. He had just returned from many years in Singapore. Rosemary and John had spent many years together cruising the canals and rivers of England, bee keeping, archaeology and many other interests and in retrospect, I think he felt her loss deeply, as he died four years later.

Jill and I trailered the E9 Scott/Norton mark II up to Abbottsholme this year, fitted with the phased transfer manifold as a static display, to show that we have, at least, built what we said we would four years ago at the lecture!

Second holiday this year, in October was to Thornham in Norfolk, with Charlotte, it was to be our last real holiday with friends, as Jill was becoming more forgetful and frailer, she had great difficulty in remembering where the cutlery/crockery etc was each day, having to be repeatedly told. This is apparently the way things go with this type of illness; once out of their familiar surroundings, they become lost! Awful. That's why it's best, if possible, to keep them at home for as long as it can be managed, requiring 24-hour supervision where necessary. Luckily, we were able to do this; my health being reasonable, and I was with her right up to within five hours of her death!

Our digs here were an old cottage on the High Street, amazingly comfortable and well appointed, opposite the road that led to Thornham Harbour on the coast, connecting it with the Norfolk coast path leading to a bird sanctuary near Brancaster Staithe, that Charlotte walked to, the day Jill and I drove to Sandringham. Trips were made to Holme, next village west of us, a working mill at Great Bircham, Hunstanton to view the 'red' cliffs etc and return to Holkham Hall, where, at the motor museum, they had a sports car special with a 2CV French engined, flat twin cylinders, nicely done (ref photo.)

We ate out several times, usually at the 'Gin Trap' at Ringstead, which was only about three miles from our digs and was very cosy. On our only trip to Wells-next-the-Sea, for a fish and chip lunch, as we were strolling down the promenade, there was a 'coarse' shout from a passing motorist, who swerved into the curb shouting 'Collett', it turned out to be 'Fish', one of Steve's motorcycling mates from Bedfordshire who now lives near here in Norfolk and was out shopping with his wife and mother-in-law! He unfortunately didn't have time for a pint! Terrible. Roger Moss paid a visit in October to view progress and the possibility of a road run, but the engine was uncooperative, refusing to run evenly on all four cylinders, this was before the new carburettors were fitted.

I received a letter from Hunting's Personnel Department to say that they had received a communication from a Mr Cummings he was preparing to write a book about design work for the MoD (Ministry of Defence) and was therefore contacting various firms involved. My name had come up, presumably due to my patents etc but so far, they had withheld giving him my contact details, in case I was not interested. Anyway, I contacted him; the net result was that he came and paid a visit here and we discussed all the relevant details of all the projects I have previously mentioned, plus some others. He said that I had been the most productive and the most helpful of all the people he had seen so far and as a result took away almost all of my relevant paperwork, for further evaluation, I have kept in contact with him ever since and now 2019 understand that it is to be sent to the archives of the RAF museum at Brookland's.

Finally, on a more depressing note, my friend and helpful colleague Brian Herbert died in December at the age of 75 after five long years of Alzheimer's disease, during which time he was nursed by his wife and son. He went to the Nuneaton crematorium in a motorcycle and sidecar, which was attended by a contingent from the BSA owners club!

2014 started off badly, with the news that John Lane had gone into St Thomas's hospital in London for major surgery for an aorta aneurysm, which lasted for about three weeks before being sent home, where he made a most remarkable recovery. It is thanks to Lyn mainly that he is here today, with all the problems he has had over the years, she has stuck by him all through – that's true love!

Jack's funeral took place at the Vale in Luton on Wednesday 16 April after a short spell in the south wing of Bedford hospital, where he was taken after being found at the bottom of his stairs one morning after the neighbours became alarmed at not seeing his curtains drawn back in the morning. He, like Brian Herbert, was driven from home in a motorcycle combination with a motorcycle escort including Steve and Shed, both on their 'Black Pigs' i.e. 1200cc Royal Enfields of 1938 vintage – very impressive!

There was a fly over Flitton by a Spitfire. Flown, I understand, by a woman pilot and the RAF had sent a staff Sergeant in full dress uniform to stand at attention in the Stopsley crematorium entrance and the wake was attended by everybody including some members of the old Vauxhall UFO group, which was the last time I saw Derek Keep. I think of him constantly and of all the good times we had together, he went on more than 40-odd holidays with Jill and I over the years and expanded my knowledge immeasurably – Thank You, Jack. Our Holidays this year, consisted mainly of visits to Lyn and John's, at Easter, Aylesford and New Year, where we tried to be of some use whilst there!

The first road test of the E9 took place in June and was a 30-mile run around the local area, culminating in a seizure, momentarily on Clickers Way, Earl Shilton and a four-mile push home as it wouldn't start again – oiled plugs! It was in the phase transfer mode and went very well, completing about 30 miles, before being pushed home on a hot day by a 79-year-old man! It's a heavy bike; I had a sleep halfway home, under a bush next to the M69. This was all covered in the December issue of *Yowl* by which time it had covered 344 miles on the road without too many major problems! A word of explanation for those of you here who will not read this issue of the *Yowl* and are wondering whether or not it is better than a normal two stroke. I quote:

"In the normal mode, the starting appears easier, probably due to the crankcase pressure forcing the mixture in better at low revs, rather than just piston displacement. However, the tick-over is less stable than phased transfer but if the revs are allowed to drop too low, it will cut out (carburation, ignition, etc, is exactly the same for both modes) the phased transfer engine appears to have a more progressive throttle response, i.e. a flatter torque curve, whereas the normal engine seems to require greater throttle opening but the power surges in, as the revs build up, with the possibility that it might be more powerful right at the top end but as I am still running in; only 2/3 of a full throttle have been used, the fastest speed so far being limited to 75 mph, which is 3,000 rpm in top gear and it pulls a 3:1 gearing easily. The fuel consumption in the phase transfer mode appears to be slightly better than normal. From my rough measurements, I would say that phased transfer gives about 35 miles to the gallon and normal about 30 miles to the gallon, a lot of this is on country roads in the intermediate gears and would undoubtedly be improved on major roads, running in top gear at a constant speed, as is given in ordinary road test data."

The bike is now off the road, for modifications to the coolant system, lubrication system and the fitting of the twin leading shoe front brake, plus stronger fork springs. Hopefully then, we will be in a position to put it on a dynamometer for accurate, power, torque and fuel consumption figures. This has not happened yet, March 2020, I'm awaiting completion of Steve Linsdell's new dynamometer facility at Flitwick.

Finally, this year we had a good 'Beano' for my 80th birthday party at The Bull pub in Stanton (ref photos). With a large attendance, which pleasingly contained John Collett. Also, this year we had a new steel roof fitted on the three garages, as the old asbestos one had been cracked by young boys playing football in Abbotts Yard, and of course, the removal of the asbestos was a problem, solved by money, as usual!

2015. The E6 has been standing since 2011 on a SORN and has effectively seized up, so a strip down was in order. We suspected that the ethanol in modern petrol was the answer, as we had to make a mandrell and knock the sleeve out of the cylinder bore as it had corroded and welded itself together. This effectively brings the saga of the homemade sleeve valve engines to a close after a span of 42 years from E1 to E6, i.e. 1973 to 2015.

To sum up: The E2 four stroke did 5,808 miles in total.

The E3, 4, 5, 6 two strokes 5,223 miles in total making 11,031 miles the total for all home-built sleeve valve engines.

After discussion with Fred it was decided to offer the E6 machine, now restored, plus the original Scott/Norton racer, now back from the Donington museum to the Sammy Miller museum or the National Motorcycle museum, Birmingham. Both have refused saying they are at full capacity at the moment but if the situation should change, they will contact us – don't call us, we'll call you!

It was now decided, as Jill wasn't well enough to go on the back of the Rudge, Jack had died and I was riding the E9, and about to become the owner of John Lane's Dominator; that the Rudge should go to Miles, this had been discussed with Jack previously with his agreement. So, I trailered it down to his house in Bedford, with Jacks old dispatch riders coat (ref photos). I resigned from the Rudge club and joined the Scott and the Norton clubs instead, this being more appropriate for me! Scott/Nortons, one and two.

The second Forest Drive reunion came next in July; this time organised by Bob Durrant at the Five Bells pub in Chelsfield near where he lives! And later at the nearby vicarage garden for tea and cakes, lovely! The original 13 was now reduced to 10, Mick Davis declined to come. Colin Shields had had one or two minor strokes and Clive Heasman's health was not up for a long trip from Devises. The weather was good, and a great time was had by all, with talk of another – whenever! Lyn Tindall made a beautiful cake – another of her gems!

Now that the Rudge had gone, I had to have another bike, don't ask me why? Fred thought it was crazy in view of the work in hand, but I suppose the desire to ride was still with me. So, I made an offer to take all of the twin cylinder Norton bits, which included two and a half spare engines, one of which I had given him previously and some spare bits and pieces, all for £4,000. Offer accepted, I trailered it all home and Fred and I started restoration work, it had been standing since 1999 as you may remember.

Later, I helped John to remove the remains of the solar heating system he had installed in the 1990s with success and all that copper reached £120 at the scrap yard. Miles and I trailered the E9 up to Abbotsholme for the last time as the new headmaster is not sympathetic to the cause and a new venue is required! What a shame we started it up and I rode it around the playing field perimeter, answered many questions, shook hands with Roger Moss to almost conclude a successful project without whose help, Fred and I would never have completed same! It has now done a total of 378 miles! The main photograph on the October issue of the *Yowl* shows me galloping along, demonstrating phased transfer! From whence it returned home to be stripped down for further development modifications mainly to the coolant system, which was still being troublesome.

In September, the Lane's celebrated their 50th wedding anniversary with a round Tour of Britain i.e. England, Scotland and Wales in that order, in a car with a slipping clutch, that Lyn did not know about! They went up the east coast to Newcastle to see her relatives, on up to Altnaharra in Sutherland for a stay of two nights, which cost as much as the rest of the holiday. Then a tour of the eastern Scottish coast using bed and breakfasts, crossing over to Morecambe and returning via Ron and Monty's in mid Wales, thence home. John Collett tells me he has now sold his canal narrowboat, after owning it for 14-years during which he travelled most of England's waterway system with Rosemary!

And finally, Jill's 80th birthday at The Bull pub again with John Collett in attendance, plus fellow nurses Charlotte, Jean and the rest of the family.

2016. We were invited to the opening ceremony of a memorial barn on Flitton Moor, a tribute by the villagers of Flitton, for all the help Jack had given over the years. Miles and Shed turned up on Jacks old bikes, i.e. the Rudge and Velocette (Dottie) respectively, it was nice to see everyone again and I understand that it may well become an annual event; goodness knows what Jack would have thought of it, he would have curled up with embarrassment! An exhibition of his work was on show and some could be purchased, the proceeds being donated to the village fund.

Meanwhile, work was continuing on the Dominator restoration with Fred's reluctant help, when an invitation from the VMCC arrived for a meeting at Cadwell Park, for all those competitors who had competed at the original first organised VMCC meeting there on Saturday 21 May 1966! Well, of course we had to be there and Miles was keen to go, so Fred and I knuckled down to work on the original Scott/Norton. It had not run for some time having been in Stanford Hall and Donnington museums for many years, also since restoration work by Roy Shearward in 1997.

We had two nights in a bed and breakfast at a small village near the circuit with Jenny; and I rode around for three laps before retiring with slipping ignition timing, a bit disappointing but my own fault, I forgot to renew the araldite holding the Magneto drive sprocket together! however we met many old friends, including Roger Moss of course, Ted Snook who ended up suffering from dehydration due to the hot weather, Simon Gregson without his Norton – shame and Jack Squirrel, over from Switzerland where he now lives and incidentally pleased with our Brexit result!

As a result, my original 1964 article written for the Scott club was reprinted in the *Sunbeam Motorcycle Club News* – October/ November issue, as well as photos in the present Scott magazine, *Yowl*. We completed the Dominator and having registered it in my name etc, I rode it around and then took it to the Norton owners club meeting at Welford Wharf, where it was well received, being the oldest machine there and more or less the only one with a kick start! Seems they have all got, leg problems and their only in their sixties!

Work now resumed on the E9's water system. the trouble being eventually traced to leaking cylinder head gaskets and their eventual remedy, all fully described in the notes and a later February 2019 issue of *Yowl* delayed, because Roger Moss 'lost' my notes for a year! During most of this time with the E9, I had been giving thought as to how it could be simplified and turned into a more commercial proposition; no self-respecting engine should have more than one crankshaft, so the idea of alternative fours was born. There are several alternative arrangements – VIS.

1. An inline four, but linking the end cylinders?
2. A radial four, difficult installation problems for automotive uses – suitable for aircraft perhaps?
3. Flat 4, reasonable solution.
4. Vee 4, most likely, does not necessarily have to be at 90 degrees, the most compact arrangement, is at narrow angles with crankpins arranged to give 90-degree firings.

All this now resides in the E10 folder and no large-scale drawing have been done! For reasons that will emerge as you read on!

Finally, towards the end of the year, Fred and I began to hear child-like voices and screams from the adjacent garden, to realise that we had new neighbours.

2017. Our last diesel car finally came to an end of its life after 10 years of faithful service and 150,000 miles, to be replaced by a petrol engine Vauxhall Zafira, back to the old firm! The government, in its wisdom has decided after years of support, that diesels are more polluting than petrol! Read all about it! Final road trials of the E9 around my test route were conducted with the new cylinder head 'O' ring gaskets in April, and eventually reported in February's 2019 *Yowl* magazine.

A tragic letter from Pat, Colin Shields sister, informed us of his death at the age of 84, one of the original Foresters Drive seat, mob.

The arrival of our new neighbours, with their two little girls at Beauly Cottage, caused the change in direction of my engine developments. The gentleman in question, a Mark Warrender, turned out to be a motorcyclist of some renown; he came back from Australia, over land, by himself on a motorcycle. No mean feat! Discussions with him revealed that he had bought the house for the size of the garage as he intended to turn it into a workshop! Furthermore, he wanted to build a large single cylinder 'sleeve valve' engine by grafting a Bristol Hercules cylinder, piston and sleeve assembly, onto his homemade lower half; similar to bikes like the Flying Millyard etc. This was a huge Vee Twin with two poppet valve cylinders off a Pratt and Whitney, American aeroengine, that was currently doing the rounds! Well, he couldn't have come to a better place everyone said – talk about Providence, astonishing! Also, I had all the original Bristol drawings from the 1940s/50s for the Hercules engine, plus a maintenance manual giving all technical details! So, the E11 was born, in July when the

GA – General Arrangements drawing was finished, 'Project Hercules' was born. Meanwhile, Mark was collecting the necessary Hercules bits and pieces and on one occasion, I went with him and his father to an aircraft museum near Newark where they generously gave him some items, upon hearing of his intentions and we started work in earnest, by machining the con rod from solid.

The man who bought John Lane's CSI Norton racer, has now rebuilt it and was demonstrating it at various venues, including a trip to Montlhery race circuit in France (ref photo) and John himself was called upon to find and display all of the trophies he had won over the years on his various CSI's – (ref photos).

My cousin John's death in November was not totally unexpected as he had been suffering from a genetic heart condition for many years, unbeknown to me, I thought he was the sort of chap who would live forever! I read a tribute to him at Saint Oswald's church, Norbury which he had attended as a child and he was buried in Bandon Hill Cemetery, Beddington in the family grave. Then we all retired to the good Old Plough Inn where we had played snooker, many years previously. I did an obituary for him in the old Whitgiftian's news a few weeks later.

Coming home with Jill from shopping one morning, we were accosted by our neighbour, Penny Butler (nee Smith) who asked if we would sell her the green lane behind her house as she had no back garden or access without permission from either Jill or me. This was the start of our downsizing programme that was to continue throughout the next two years during which we sold off, the workshop and rear garden, plus number 51, our rented cottage next door, that our tenants had decided to vacate as they needed a larger house to accommodate their ageing father.

2018. Ann Roach (nee Symes) died after having had Alzheimer's for several years and nursed at home by Barry, bedridden until the end. We had some good times together with Richard, she was a livewire!

Work continued on the E11 as material arrived for the flywheels, main shafts and sleeve shaft assembly etc all through the year.

Ted snook's 80th birthday was held at a pub in Newport Pagnell, attended by Miles, John Urwin and myself. Unfortunately, Jill was now not up to the effort and stayed at home with a carer, as we were now having to bring in support from 'Salus', a private agency who came in the morning to get her up, wash and dress her, whilst I prepared the breakfast etc and then later I was able to get the meals etc until bedtime, when I settled her down for the night.

Terrible news, Fred died of a massive heart attack, within 24-hours. I had spoken to him about a week before and he didn't sound at all well, he said he thought it was flu! His wife Jeanette was also developing the Alzheimer's disease and he had been helping her with the shopping etc, the burden of course, has fallen on Paul, his only child, who I understand has moved his mother closer to home! End of an era! No Fred – without his help the E11 will slow down. Rest in Peace, Old Friend.

On a more cheerful note, Jill and I were preparing for our silver wedding anniversary on 6 March. However, her aunt Pat died at the age of 91, she had been a good friend for many years and we used to call round almost every week after shopping. Coinciding at this time, was the arrival of the 'Beast from the East', a week of extremely cold weather from Siberia that paralysed the country; snow everywhere, we burnt more logs that week than we had in the last 10 years, I think! It reduced the silver wedding guests by almost 30%, sod's law! The sale of 51 was completed on 12 April and a builder from Romford and his wife took residence, it was a relief as we had been paying council tax, electric and gas bills for three months over this winter!

Bob Durrant dies, 84; it is becoming more like an obituary than my memoirs; I think it was cancer, he was in hospital for a week or two before and had phoned his wife Margaret saying, 'Get Me Out of here, they're trying to kill me' and so it came to pass! I am sure they did their best.

We went to the Pailton LE Velocette owners rally with Ted Snook who helps to organise it and had a ride on the steam railway around the garden and the steam launch around the lake, all very pleasant, the weather played a ball.

Next up is my sister Phyllis, aged 81, a merciful relief for everybody, I had been hoping for this for several years as she was totally out of this world and didn't know anybody, no conversation, no recognition whatsoever, dreadful, nobody should have to endure that! I am glad my parents never knew, poor Phil she got a rough deal from life! We attended her funeral on the outskirts of Wellingborough and the wake following; I gave a tribute to her, mainly talking about our early days in Baildon and Wallington leaving Peter and Paul to concentrate on her latter days.

Tribute to Phil.

> *"She was a quiet girl rather like Dad and I think they had a lot in common. He used to help her as much as possible, Mother was a bit impatient with her at times, if she was too slow or did not show much enthusiasm, like Dad, for an idea. She nevertheless had a lovely temperament, slow to anger but could stand up for herself when required. Never seemed to expect much out of life and just accepted things as they were; unlike a lot of people today!"*

The sale of the workshop and rear garden to Mark and Jenny Warrinder, completed our downsizing programme to just three houses and three garages! With this, I have now, in effect, become a 'Fred' working for Mark on his own 'Project Hercules'. What a joke, I wonder what Fred would have said?

Iris's 80th birthday took place at the Brasserie restaurant in Woburn, organised by Sarah with jazz accompaniment. We went down to John and Lyn's, for the last time with Jill, as it turned out, in November, taking the wheelchair and commode etc. They have always given great support, for without them, we would not have had any holidays these last three years! Finally, we had a visit from Cousin Martin Rush to collect his theodolite that I had fixed to a set of tripod legs, he had purchased elsewhere. He came with Cousin Jill and Andrew, his sister and nephew, respectively. Richard and Jacqueline came as usual with Miles later for Christmas.

2019. This year started very badly with the death of Jill, just about a week after visiting Whitemore's antique centre, where we met Mark with his parents. She collapsed one morning whilst the carer was getting her dressed etc., and we thought she'd had a stroke and called the doctor, who finally arrived late afternoon. After various tests she called for an ambulance which arrived early evening and took her to the George Elliot hospital in Nuneaton, where she used to go once a year to visit the consultant. I did not go with her, much to my regret but was exhausted both from stress and lack of sleep, as I had had some consecutive bad nights with her.

I got a call from them early next morning telling me to go there as soon as possible, which is what I was intending to do anyway, so after phoning Jacqueline and Paul, I went up with Mick. The Doctor there, took us aside and said she was not responding to treatment and did I want to intensify it. We both agreed on 'no', her quality of life was already very poor and to prolong the suffering would have been cruel. She died later that afternoon with Jacqueline, Richard and Paul at her side. Mick and I said our goodbyes that morning, but I wish I had returned to be with them!

It was a terrible shock as she seemed to go down so quickly at the end, dying of pneumonia, so the death certificate says. But it was for the best really as she had suffered enough and she was still Jill. We all miss her terribly of course, she had so many friends, evidenced by the big turnout at her funeral, about 100 people passed through with the wake as well. Still, as Charlotte said, 'she did it her way,' and in doing, gave me back my own life. Dearest Jill, she was 'my friend'.

It has taken me a long time to write these last paragraphs and I know we all think of her, every day. The funeral took place two weeks later at the Heart of England crematorium, after a visit to the chapel of rest for further goodbyes! Followed by the wake at Barwell Bowling Club, where over 80 guests attended. I would like once again to thank everyone who came and everybody who helped to get us through this very, very difficult period!

The weather was superb in February of this year and after the funeral, Alan who had, of course, come over from Canada, Holly, Sarah and I took a walk over Croft Hill in glorious sunshine and met Joe Vernon and husband Chris, and dog, also walking over the tops. Also, a rather unusual thing happened a few days before she died: she suddenly requested me to take her down to visit Mick and Dot one afternoon, so I wheeled her down in the wheelchair. This was most unusual as at this time, she was more inclined to stay in – it makes you wonder!

Following all this was the start of the legal process, to obtain probate etc further complicated by having had to pay capital gains tax on the sale of number 51, for which I employed Integra Accounting Ltd of Hinckley and the family solicitor, Flavells Ltd. Even so, it took almost the best part of the year to finally bring it to conclusion. Another funeral in March, for my 92-year-old cousin Gill with the 'G', took place at the Warwick crematorium, the last time we saw her was with Martin for his theodolite collection last year, remember?

Next up was a trip with Sarah, to Baildon for the scattering of Phyllis's ashes, with Miles, Jenny, Peter and Paul. Sarah and I went up on Friday afternoon, to meet Miles and Jenny who were coming up the A1, for a farmhouse bed and breakfast at Shipley Glen. We arrived slightly before them as they had been delayed, when a lorry driven by a Polish chap, drove into the side of them, somewhat denting the car but not immobilising it. As usual he came out alright. Luckily, the insurance company wrote off the car, paid him out and then he bought it back for peanuts and is still running it – he hasn't knocked out the dents as he says it keeps people away, they steer clear of him! It's not an MOT failure problem unless sharp edges are involved!

The idea, apart from scattering the ashes was to show them, as much as possible, the places of interest that Phyllis and I were familiar with as children. So naturally, we went to 30 Ferncliffe Drive with Peter who had digs in Shipley, Paul was to come up the following day – Sunday. The people who lived there now were very good and had agreed to us visiting and looking around the house and garden and scattering Phyllis's ashes in the garden, which we did with Paul the following day.

We now went down Baildon Bank, as Phyllis and I did to visit Sandals Road School at the bottom, which is now converted into flats! What a shame. Then along the line of cliffs to the quarry, where us lads played in days of yore – what memories. We also went to the Cow and Calf rocks on Ilkley Moor that Jill and I had visited many years before. We had lunch at Dick Hudson's pub on Rombalds Moor and then onto the famous Bingley locks, a flight of five, very steep and finally Baildon station where my dad used to go to work in Bradford, which is still running, although now reduced to one platform and single track!

Now came a trip to Otley to see David Berry my oldest friend, I have known him for nearly 80 years! This is said in 2020, astonishing! He and his wife Pearline are in reasonable but shaky health, it's the first time that they have met Miles and Sarah. Finally, all six of us retired to Shipley Glen where Peter scattered the remaining ashes and we all took a trip on the Glen cable railway that now runs at weekends in the spring/summer, Sarah and I returned after this on Sunday afternoon, leaving the others to return on Monday.

Daffodil Sunday came and went with Richard, Jacqueline, Charlotte and Jean, as usual, followed by visits to Whitemore's, where I was looking at a display of tools for sale. A gentlemen, upon seeing my Scott/Norton badges on my jacket said that he had once owned a Scott and from this it led on to two strokes and Trojans! Surprise, Surprise! Well, the long and short of it, was that he had a book called *My Trojan Story* by a Mick Reed, ex-apprentice that I knew! He said he would post it to me, which he duly did! It's a small world.

As a result of this, I joined the Trojan Museum Trust, sent them a selection of Trojan memories, talked on the phone to Mr Reed and now receive their news sheet twice a year! I finally put all my road going motorcycles on a SORN as both the road tax and insurances came up for renewal, they were now getting a bit heavy for me, the Dominator to kick over and the E9 Scott/Norton, just too heavy to lug about! Although it has an electric start.

Another visit to John and Lyn's in May resulted in my clearing up the remains of their swimming pool which had been lying about for years and hiding it out of sight behind their top garden sheds, it has now become a vegetable garden area! Upon my return to Stanton, I was off again, this time to John Collets farm at South Nutfield to arrange the pick-up of various machine tools, to go in Mark's garage. All this was a result of discussions with Miles and Mr Ken Hamilton who is looking after John's interests, until probate is granted. They suggested we get them away before the handover to the new owners, 'The Countryside Trust'.

So, down we went in Marks firm's van as we had done previously, to collect all the smaller items, with Igor as our driver, one of Mark's staff, to get everything ready for the pre-arranged arrival of the lorry with crane, the following day. We were to bring back the Bridgeport Mill, The Jones and Shipman 540 Surface Grinder and Myford Lathe. We actually bought back all these, plus a two-speed pillar drill, a table linisher and many more machine tools. The weather was terrible with thunderstorms but nevertheless we managed to get it all home, dry, and in one piece.

The Big Apple tree now in Mark's part of the garden, fell over, completely missing the workshop, Penny Butler's new fence, and the existing hedge that used to belong to John Vernon – unbelievable. He is going to chop it up for logs now that he has installed a log burning stove!

I spent the last week in June on the North York Moors with Steve, Shed, Pete Wheeler, Mervyn, Bondi and two others, we were doing a sort of mini Three Tors, as of yesteryear! I was invited to ride Steve's 500cc 1938 Royal Enfield known as 'Piglet', whilst most of them rode their Black Pigs, 1200cc 1938 Royal Enfield's with an International 350cc Norton and a 500cc 1938 Rudge. All in all, eight machines all pre-war, great to see! I was picked up here at home and went with Don Palmer and Fish in the Flitwick motorcycles van, as backup! This proved its worth on the way there, when Shed's Black Pig started to smoke and was deemed too sick to continue, just south of the Humber bridge.

The place we stayed at in East Ayton near Scarborough was great, it had apparently been a reform school in Victorian times and therefore was built around a courtyard, assembly area with rooms and balconies overlooking, and standing in, its own extensive grounds, ideal for bikers. The weather wasn't good for the first half of the week, so an investigation into Shed's pig was made. It was thought, at first, that the piston rings in the rear cylinder had broken but this was not so. Whilst the engine was being rotated, oil was seen to be gushing from the interconnecting hole with the timing cover and subsequent removal of this revealed that it was completely full up with oil, which had nowhere else to go! Which explained why the oil sump was nearly empty, it was all in the engine! Steve, who understands these things, immediately started poking about and discovered a blocked drain hole and upon the restart, all was well. The mystery remained as to how it got from Flitwick to Humber bridge in the first place.

With the weather improving, a run was in order and I suggested, heading northwest across the middle of the Moors to Roseberry

Topping, at 1,046 feet to view and possibly climb its beautiful cone shape. A damp and misty ride over hill and dale took place, no climb, and we returned over the moors for tea and cakes at Danby, then Goathland, where the TV programme *Heartbeat* comes from, across the North Yorkshire Moor railway line and back via the A169.

We followed this up with a run to Pickering, to view a vintage car/ bike museum and this time with the much-improved weather, I rode Piglet. As soon as we arrived, we were seized by a young television producer who was making a film about the museum, etc and having almost immediately invited us to the nearest cafe for morning coffee etc, all expenses paid, wanted us all to appear in a forthcoming documentary, East Yorkshire television, so we signed the usual disclaimers and agreed!

They did some filming of us in the museum, looking and commenting on various vintage bikes and whilst we were preparing to depart, to continue our run. It was then that the accident happened. I had just started my machine and was sitting waiting for the others, when Bondi, who had just started his black pig, pushed it off the stand; unfortunately, it was in gear with the back wheel rotating and it literally threw itself at me! It hit me on the offside, crushing my second index finger on my right hand, the one I am now writing with, and fell over with Bondi still trying to get the clutch in.

Amazingly, there was virtually no damage to either bike and only the headlights pushed slightly to one side, it didn't even break the glass on either machine! But my finger was the worse for wear! Bleeding profusely from a deep gash and badly bruised; all on film, I believe. Somebody somewhere produced a bandage and we proceeded on our way to Whitby for a look around the River and dock area, followed by a visit to Scarborough and some laps around the Oliver's Mount Road Race Circuit, thence home.

I had a bad night with the finger, stopping the blood from getting on the bedding, when I fell asleep. So, it was decided to have it properly dressed and I was taken to Scarborough hospital in the van whilst the rest of them set off for Eden Camp heritage museum near Malton, where we joined them later. They even had a Bristol Hercules engine on display. We returned the following day via Fernando's for lunch and hope one day to see the film on TV. I believe they were going to notify Steve when it was to be shown!

Stanford Hall VMCC Founders Day gathering of the clans again with Roger Cramp, Miles, Jenny to meet Roger Moss, Simon and Sandy Grigson, Sheila Neal, Mervyn, John Hurlstone and Spike, who wants to buy the Dominator! I gave them a trumpet demonstration from Hurlestone's market stall (Ref Miles!)

On the 24th of July, Dave Tebbut died of Parkinson's disease at the age of 81. He had been a good friend and colleague for many years and of course, we enjoyed many holidays together. His mountaineering exploits alone would make very interesting reading and I believe Sandra is now attempting the challenge! I should like to read them eventually.

Another visit to VMCC Aylesford, no Dereck this year, I think he is too frail, to meet Bob and Lyn Hamilton, Simon and Sandy Grigson, Miles and Jenny. Followed by Pailton, without Ted Snook, he was also too frail – a sign of the times!

Nick Moss died this month, August, age 66, only just retired, a great blow for the E11 project as he was just about to start making the patterns for the Hercules engine, we are still looking for another but it seems they are dying off quicker than we can get hold of them!

Jill's final internment at the Saint Mary's church graveyard was very much a family affair, with all her brothers; Alan from Canada with Holly, Miles, Jacqueline, Richard, Paul, etc where her ashes were put alongside Ray and Matthew, leaving space on the new tombstone for me, when the time comes!

I went down to the Lane's on an open-ended stay, to help Lyn after John's knee operation, which was now imminent but first, we attended Gareth's 50th birthday party at Thames Ditton, at a pub on the River Thames, opposite Hampton Court Gardens and afterwards at their home in East Molesey. John's operation took place the following week and he was kept in hospital for about four days. During this time with him out of the way, and with Lyn's permission, I undertook the challenging task of clearing the garage and removing all the valuable CSI Norton items I could find scattered around in the various sheds; to the safety of the dry garage. It was partly inspired by an enquiry via Simon Grigson, from a man in Bristol who wanted some clutch components.

When all was done, I estimated that, apart from a frame and oil tank, there was nearly enough to make up another bike! So, he has still got virtually two complete CSI's, having already sold two and possibly a third in bits. No wonder Rex Boyer said they were his pension; I doubt he paid more than £100.00 for the lot! Upon his return from the hospital, we waited with some trepidation but he was 'delighted' with the result! Although he was unable to view it all directly and had to be satisfied with pictures off the mobile phone.

As a treat, I was taken by Lyn and Teresa to their ukulele class in the local village hall, which they attend once a week, to hear a series of recitals, some of George Formby's etc, they were very good to my untrained ear and I enjoyed about two hours of entertainment! Plus, tea and cakes.

My cousin Martin Rush was now 80 and gave a party to celebrate at a pub in Bridport, Dorset for which Miles drove Jenny and myself down there in pouring rain, one Friday afternoon to stay for two nights at a bed and breakfast, run by two gay gentlemen. It was a good do and lasted till about one o'clock in the morning – still raining, in fact it rained all weekend, terrible weather. I hope it's better next year when we are supposed to be doing another motorcycle Three Tors type rally in this area!

Another final visit to John and Lyn's for my 85th birthday, saw me finishing the rubbish clearing up and sorting out his garden workshop; thence onto Christmas with Miles, Iris and Peter, as Sarah had gone off to Trinidad as part of Gareth's 50th birthday celebrations, therefore unable to have Iris, and no Jacqueline and Richard, as she was ill.

2020 coronavirus arrives?

1145 HP Merlin engine at Cnocbreac, Jura - June 2000.

Caves near Corryvrechan whirlpool in North Jura. Sandra, Dave and Jill.

Napton on the Hill - April 2000. Jill, Lyn, Bob Durrant,
Jack, Rob and Margaret Durrant.

'Design office' in Stoney Stanton - 1994 to 2021 so far.

E3 sleeve valve 2-stroke 500cc - April 2001.

Reliant 3-wheeler van. Jack hoping to sketch at Portpatrick,
Rhins of Galloway – September 2001.

Summit of Ben More at 3147 feet on Mull. The Achioch
ridge looking east -September 2002.

Fingal's cave on the Isle of Staffa with Jill - 2002.

Jill at Devil's bridge - Wales. Kinnerton holiday - May 2002.

Summit of Ben Scrien at 1251 feet on Eriskay. Dave, Sandra and Jill.
'Whisky Galore' film was made here.

3 tors! Lunch stop near Deeping Saint Nicholas A1175.
L to R Steve, Jack, Derek, John and Peter Wheeler - July 2002.

Scott owner's rally in the Falls of Measach, Ullapool with Jack.

Jill and Jack at Bettyhill, Sutherland - May 2003.

'Eating again'. Jack, Jill and me at Ingleby Greenhaw,
North Yorkshire Moors - 2003.

'The waitress' - Jill at the tea gardens of Huntington Hall - June 2003.

Rievelaux Abbey – Charlotte, Jill and me.

Tom, Sylvia, Jill, John, Lyn and me. Jack took the photo – Sligarn, Isle of Skye -September 2003.

Summit of Glamaig looking east towards Raasay and Applecross - 2003.

John and Rosemary's canal boat 'Magdalen' at Husbands Bosworth.
L to R John Collett, Jill, Phyllis and Rosemary - August 2003.

Isle of Arran - L to R Clive Heasman, Rob, Jack,
Mary Heasman and Jean - 2004.

Lunch stop near Tregaron, Wales. John, Rob and Jill - 2005.

Jill and I – Hay Bluff.

E4 'Shnurle' ported, 500cc 2-stroke with variable port timing.

Barry Roach's 70th birthday at Coulsdon Court Hotel.
L to R Jill, John, Rob, Linda and Rod Evans.

Edna's back garden – Croglin Eden Valley - April 2005.
L to R Sandra, Edna, Dave, Jack, Norman and Jill.

John Urwin and Miles in Milton, South Uist. Arrived on bicycles - April 2005.

Jill on summit of Hecla looking west towards rocket firing range, South Uist.

Rocket launch from Geirinis, South Uist - 30 degrees elevation - April 2005.

Jill and Jean on the road to Carsaig, Isle of Mull - September 2005.

Ironbridge - February 2006. Jack, Jill and Charlotte behaving badly!

At the Tebbut's house – Ivegill, Lake District. Jill and
Rob on Jack's 1939 Rudge Ulster.

406

E5 radial ported 500cc 2-stroke with variable port timings -
note hand operated lever.

John and Annie Urwin at Jack's house, Flitton in their 1926 Morgan three-
wheeler with Blackbourne engine 1000cc.

King Williams College, Isle of Man. Steve testing the 'Paton' 500cc.

Summit of Snaefell at 2015 feet - Isle of Man - 2006.

Myra and Jill at Pentre Ifan ruins, South Wales - September 2006.

Summit of Carningli at 1010 feet. Jill, Ted and Myra Snook.

High force, Teesdale, Durham - April 2007. Jill, Rob, Charlotte and Jean.

John and Steve working on the 'Paton' at King Williams
College for the Manx Grand Prix - August 2007.

E6 sleeve valve, uniflow 2-stroke, 500cc - March 2007.

Mire Garth, Dentdale, Yorkshire. Richard, Jack, Jacqueline
and Jill - September 2007.

Summit of Pen-y-Ghent at 2252 feet. 'Project Richard.'

E7 diesel 500cc 2-stroke not completed yet; requires motorcycle to finish!

Foxton locks with Jill and Valerie Morris, an old friend - September 2007.

Jean, Charlotte and Jill at Fernworthy reservoir, Devon.

Summit of Buttern Hill at 1345 feet - October 2008.

Sadie, Jack, Jill and Helen in Dawlish park, South Devon - 2008.

Tan Hill, Yorkshire. L to R Gav, Jacqueline, Jack,
Jill, Trish, Pat, Tommy - November 2008.

At John Easten's house, Tilchmarsh - August 2009.
On guitar -Mike Jones, banjo -John Easten; both from Huntings.

Scott rally at Abbotsholme - August 2009. 'I don't know what I've just said to her?'

Charlotte, Jean and Jill on the 'Bosworth' at Napton - October 2009.

Earl Sterndale, Peak District. Jean, Charlotte, Richard,
Jill and Jacqueline - lunch stop.

Titch Allen's Memorial Day at Mallory Park - May 2010. Me on Jack's Rudge.

With my beloved in Jack's garden - Spring 2010.

Mick's orchard - August 2010. L to R Sadie, Helen, Martha,
Paul, Jill, Leo, Sarah, Alex, Millie, Jacqueline and me.

Welshpool station, Wales. L to R Jill, Charlotte, Richard
and Jacqueline - September 2010.

Bob Ward's Bentley with Iris, Anna, Jill and Jack at Jack's house,
Brook Lane, Flitton -June 2011.

E9 4-cylinder Scott showing 'normal' manifolds.

E9 4-cylinder Scott showing 'phased transfer' manifolds.

Jean, Jill and Charlotte having lunch near Aberaeron, mid Wales.
'Took some mushrooms home for breakfast' - October 2011.

The 'seat' at Foresters Drive in 2019.

First Foresters Drive reunion. Back row L to R Carol Roach, Geoff Gunter, Rob Collett, Bob Durrant, Mick Davis, John Lane, Colin Shields and Barry Roach. Front row L to R Anita Chaffey, Bobby Hamilton, Mike Lane, Clive Heasman and Chris Hayward - August 2011.

John, me, Janet, Jill and Anita taken in January 2012 when Janet came over from Australia.

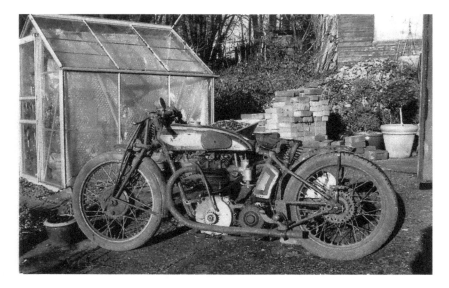

John's Norton racer rescued from the garden sheds
to a nice warm roofed garage -January 2012.

Rob, Derek, John and Jill in Looe, South Cornwall - October 2012.

Millie and Leo on Jack's Velocette 'Dotty' - 2013.

Hunstanton cliffs, Norfolk. Jill and Charlotte looking cold - October 2013.

John in a snow drift - unusual weather, Isle of Wight - March 2013.

'Daffodil Sunday' at Newnham Paddox. These gates are the
second most impressive after St James's Palace, London.

E9 1200cc 4-cylinder Scott based 2-stroke using 'phased transfer' circa 2014.

Second Foresters Drive reunion. L to R Jill Collett, Mike Lane,
John Lane, Geoff Gunter, Bob Durrant, Rachel Karn, Anita Karn,
Carol Roach, Guy Farage, Barrie Roche, Margaret Durrant,
Bobby Hamilton, Linda Evans, Chris Hayward, Sheila Gunter,
Lyn Hamilton, Miles Collett, Robert Collett. Lyn Lane took the photo.

BOB COLLETT - SCOTT/NORTON
CADWELL 25 JUNE 2016.

SCOTT 293 MAGAZINE 2016

Jill 'resting at home' circa 2016.

John Lane and his 'trophies' for road racing,
hill climbs, grass track, sprinting, etc.

The workshop at Stoney Stanton between 1993 to the present 2021.

Peter, Paul, me, Jenny, Sarah and Miles in the back garden of 30, Ferncliffe
Drive, Baildon where we scattered Phyllis' ashes on the flower beds.

Me, David Berry, Miles and Pearl in their back
garden at Otley, Yorkshire - March 2019.

Just after Phyllis' final internment. Peter, Miles, Robert,
Sarah and Jenny at the pub on Baildon Moor.

CPSIA information can be obtained
at www.ICGtesting.com
Printed in the USA
LVHW070803150522
718581LV00031B/355/J